PENGUIN BOOKS

The Penguin Book Quiz

Praise for *The Penguin Book Quiz*:

'It's so much fun. The perfect level for keen but not necessarily expert readers. Specific enough to make you reach back into your memory and occasionally kick yourself at not finding the answer, but not so specialized that you want to give up . . . I'll definitely be buying copies of this book as gifts' Emma Healey, author of *Elizabeth is Missing*

'Quite possibly the greatest social lubricant since the invention of alcohol. An absolute delight' John Preston, author of *A Very English Scandal*

'Few quiz questions make you smile, laugh or gasp. James Walton's always do. *The Penguin Book Quiz* is unceasingly enormous fun' Alan Connor, author of *The Joy of Quiz*

'Detailed, entertaining and wonderfully informative, a must-have for quiz aficionados and bibliophiles alike' K. M. Ashman, author of The Blood of Kings trilogy

Praise for *Sonnets, Bonnets and Bennetts*:

'The perfect gift for the sort of person (like me) who shouts the answers at *University Challenge*' Lynne Truss, author of *Eats, Shoots & Leaves*

'Fiendish, funny and endlessly surprising, James Walton has provided the perfect volume for anyone who loves books, relishes a quiz or just fancies showing off at a dinner party' Mark Billingham, author of the Tom Thorne novels

'How nice it would be if the family would sit around the Boxing Day fireside considering the questions in affable competition' *Daily Telegraph*

'A must for those who think they kno the unexpected, alongside the more se

ABOUT THE AUTHOR

James Walton has written and hosted seventeen series of BBC
Radio 4's books quiz *The Write Stuff*. He also writes questions
for BBC2's *Only Connect* and contributes a weekly quiz to
the *FT Weekend* magazine. He is the books editor of *Reader's
Digest* and reviews books for the *Spectator, The Times, Guard-
ian, Daily Mail, Daily Telegraph* and *The New York Review of
Books*. His previous books are *The Faber Book of Smoking* and
Sonnets, Bonnets and Bennetts: A Literary Quiz Book.

The Penguin Book Quiz

From The Very Hungry Caterpillar *to* Ulysses

JAMES WALTON

PENGUIN BOOKS

PENGUIN BOOKS

UK | USA | Canada | Ireland | Australia
India | New Zealand | South Africa

Penguin Books is part of the Penguin Random House group of companies
whose addresses can be found at global.penguinrandomhouse.com.

First published 2019
001

Copyright © James Walton, 2019

The moral right of the author has been asserted

The permissions on page 343 constitute an extension of this copyright page

Set in 11/13 pt Dante MT Std
Typeset by Jouve (UK), Milton Keynes
Printed and bound in Great Britain by Clays Ltd, Elcograf S.p.A.

A CIP catalogue record for this book is available from the British Library

ISBN: 978-0-241-99003-2

www.greenpenguin.co.uk

Penguin Random House is committed to a
sustainable future for our business, our readers
and our planet. This book is made from Forest
Stewardship Council® certified paper.

To Helen, Sam and Beth,
with much love

Contents

Birds and Bees: A History of Penguin and Quizzing   ix

How It Works: Guidelines to the Quizzes   xxxi

Quiz 1   1

Answers to Quiz 1   19

Quiz 2   35

Answers to Quiz 2   55

Quiz 3   71

Answers to Quiz 3   91

Quiz 4   107

Answers to Quiz 4   125

Quiz 5   141

Answers to Quiz 5   161

Quiz 6   175

Answers to Quiz 6   195

CONTENTS

Quiz 7   211

Answers to Quiz 7   229

Quiz 8   243

Answers to Quiz 8   263

Quiz 9   277

Answers to Quiz 9   295

Quiz 10   309

Answers to Quiz 10   327

Acknowledgements   341

Text permissions   343

BIRDS AND BEES:
A HISTORY OF PENGUIN
AND QUIZZING

It seems only right to start a quiz book with a quiz question. So here goes – what became part of British life first, Penguin books or quizzing?

The answer, perhaps unexpectedly, is Penguin books, which first appeared in 1935 and took off immediately. Not until three years later did Britain get its first quiz – when, just after 8.30 p.m. on 19 April 1938, the BBC's Regional Programme Northern broadcast *General Knowledge Bee*, 'a contest across the Pennines' between schoolchildren from Lancashire and Yorkshire.[1]

And 'broadcast' is a significant verb here, because, as Alan Connor makes clear in his highly recommended *The Joy of Quiz* (published, as luck would have it, by Penguin), the quiz as we know it today came into being only with the rise of radio. The Victorians famously loved their parlour games, many featuring surprising levels of

---

1 Incidentally, the word 'quiz' was *not* coined by a bloke in eighteenth-century Dublin who bet his friends he could popularise an entirely new word and so wrote 'Quiz' on walls all over the city. The story is an early and impressively long-lasting example of an urban myth.

physical contact and low-level violence.[2] Yet it never seems to have occurred to them that there was fun to be had by simply asking each other, say, 'Who was the first British Prime Minister to have a beard?'[3] or 'What connects HMS *Bounty* with the English cricket team that toured Australia in 1882–3?'[4]

Admittedly, as Connor also makes clear, the quiz as we know it today did have its precursors. As long ago as 1691, John Dunton, a London bookseller, launched the *Athenian Gazette*, which had the neat idea of asking people to send in questions. The result was like a 'Queries' column in today's newspapers, except with a team of experts – aka Dunton and his coffee-house mates – supplying the answers instead of other readers. (Not that all their answers would necessarily have passed a modern fact-checking test. Q: 'How came Monkeys first into the World?' A: 'As Man did, by the Power of God.') Plainly, this wasn't anything resembling a contemporary quiz. Nonetheless, it did help to establish the concept that discrete bits of information – particularly when set up by questions – could be a source of enjoyment in and of themselves.

This same principle underlay the journal *Notes and Queries*, founded in 1849, which also invited its readers to ask the experts – although in this case the experts were

---

2 In the disturbingly named 'Squeak, Piggy, Squeak' someone wearing a blindfold placed a pillow on the knee of one of the other players, sat on it, invited them to squeak and tried to work out who they were. 'Are You There, Moriarty?' featured two players, again blindfolded, whacking each other on the head with rolled-up newspapers.

3 Benjamin Disraeli.

4 They were both captained by a man named Bligh.

more expert, and the questions more scholarly. In 1884 the principle was even attached to the word 'quiz' when an American called Albert Plympton Southwick published the pithily titled *Quizzism and Its Key Quirks and Quibbles from Queer Quarters: A Melange of Questions in Literature, Science, History, Biography, Mythology, Philology, Geography, Etc., with Their Answers*. But this and a couple of other proto-quiz books that followed in the early twentieth century were still intended to be read rather than to be played for mildly (or not so mildly) competitive entertainment. They also failed to ignite any sort of quizzing craze.

Or, as it would have been back then, any sort of beeing craze – because when radio programmers decided in the 1930s that asking people stuff would make for a good listen, 'bee' was the preferred term. With an obvious debt to America, Britain's first radio bees tested their contestants' spelling.[5] Before long, though, they began to branch out. Many moved on to more specialised areas, like the Regional Programme Western's *Agricultural Bee*. ('Tonight, on the principle of the Spelling Bee, teams of young farmers from Somerset and Dorset will compete in answering . . . questions on everyday agricultural topics.') But then that historic cross-Pennines face-off of 19 April introduced the idea of questions on any subject at all – making 1938, in Alan Connor's ringing phrase, 'the year when British quiz began'.

By which time there were plenty of Penguin books for the contestants to consult – and not just the sixpenny novels that quickly made the company's name. The Penguin

---

5 Spelling bees had been around in America since the early nineteenth century, and the first national competition took place in 1925.

Shakespeare was launched in April 1937. A month later came the Pelican non-fiction series, which established its high-mindedness straightaway by kicking off with *The Intelligent Woman's Guide to Socialism, Capitalism, Sovietism and Fascism* by George Bernard Shaw.

The traditional story – or creation myth – of how Penguin came about takes place at Exeter Station when the company's founder Allen Lane was returning from a weekend spent with Agatha Christie[6] and her husband. Depressed by the feebleness of what was on offer at the station bookstall, he resolved there and then to create a range of intelligent, well-designed paperbacks that would cost no more than a packet of ten cigarettes.

Like most myths, this one is a matter of some debate. For a start, there are other versions, including the suspiciously Newtonian one that inspiration struck while Lane was sitting under an apple tree. More prosaically, he may just have been taking up the suggestion made by an office junior that keen readers like him would love to be able to buy decent paperbacks for sixpence. Yet what is certain (if a little less exciting) is that there was plenty of office admin involved – because, like quizzing, Penguin didn't come out of nowhere.

Allen Lane Williams was born in Bristol in 1902, the son of a surveyor at Bristol Corporation. By his own admission he 'wasn't very bright at school', but luckily he had a relative on his mother's side called John Lane who ran the

---

6 Christie later became godmother to one of the three daughters Lane had with his wife Lettice, whom he married in 1941 – in a ceremony where the guard of honour was provided by two lines of cardboard penguins.

publishing firm The Bodley Head. When Allen was sixteen, Uncle John[7] invited him to join the company, on the
rather odd condition that he and the rest of his family
drop the surname Williams and revert to his mother's
maiden name of Lane.

In the 1890s The Bodley Head had been closely con-
nected to the decadence movement, gaining such a racy
reputation that *Punch* magazine joked, 'uncleanliness is
next to Bodliness'. One of the big names it published,
for example, was Oscar Wilde – although the experience
wasn't terribly happy. Wilde annoyed Uncle John by
seducing one of the office boys. He also regarded his publishers as little more than servants – and to prove it, gave
the name Lane to the manservant in *The Importance of
Being Earnest*.

After Wilde's arrest and imprisonment, the company
decided to head in a more respectable direction – and by
the time Lane joined in 1919 it was among London's leading publishers. To his later pride, Lane started as an office
boy himself, before working his way up to the sales
department, where he forged links with booksellers that
would one day come in extremely handy. (In the early
1950s the recently knighted Lane looked back on the days
when he had 'traipsed the city streets trying to sell a few
books' and wondered if 'that wasn't about the happiest
period of my business life'.)

From there, with Uncle John increasingly seeing him
as the heir apparent, his rise through the ranks acceler-
ated. He was appointed a director in 1924, and when John

---

7 As he was known – presumably on the grounds that the more
accurate Second Cousin Twice Removed John doesn't really roll off
the tongue.

died the following year, he continued to enjoy the support of John's widow Annie, now the majority shareholder.

Which, it soon turned out, was just as well. In 1926 Lane agreed to publish *The Whispering Gallery*, a book the journalist Hesketh Pearson claimed to have ghostwritten from the shockingly outspoken memoirs of a senior diplomat. Unfortunately, on the day it came out the *Daily Mail* denounced the book as 'a scandalous fake'.[8] Even more unfortunately, the *Daily Mail* was right: Pearson had fabricated the entire thing. The book was withdrawn, but when The Bodley Head – or 'The Badly Had' as it was nicknamed – sued Pearson, it somehow managed to lose, leaving him free to sue the company back for the 'pain and suffering' he'd been caused.

Three weeks after the second trial, Annie died, leaving Lane the majority shareholder. He immediately brought in his brothers John and Dick, and also took the decision – one that he'd develop much further in his future career – to buy in books from other publishers and bring them out in cheaper editions, which at this point meant hardbacks for three shillings and sixpence. Realising that his fellow directors still regarded him with some suspicion, Lane even agreed to publish some of the more controversial Bodley Head titles, including James Joyce's *Ulysses*, at his own expense.

But by now – whether as a result of a dodgy railway bookstall, an apple tree or a smart underling – the idea of Penguin was beginning to take shape. With the rise of radio and cinema, the British book trade felt (not for the first or last time) under attack from new technology and

---

8 The *Observer* went with the comparatively mild 'a reeking compost of garbage'.

at a 1934 conference, attended by 48 publishers and book-sellers, Lane announced that sixpenny paperbacks would be his own answer to the challenge. 'Forty-seven of [the delegates] went away on the Monday morning with no fur-ther thought on the subject,' the bookseller Basil Blackwell later wrote. 'The 48th was Allen Lane.'

Lane's motives, as he said himself, were 'both mission-ary and mercenary'. On the one hand, he wanted to make a stand against the notion that 'the majority of people are stupid, interested only in entertainment that enables them to escape from their environment'. On the other – and precisely because this notion was so misguided – he believed that 'for several years the book trade has been sitting on a gold mine and not known it . . . People want good books, and they are willing, even anxious, to *buy* them if they are presented in a straightforward, intelli-gent manner at a cheap price.'

Despite another strangely persistent myth, Allen Lane didn't invent paperbacks. In fact, they'd been around since the early sixteenth century. What he did do was publish them with unusual care – unusual, but by no means unique. An obvious and acknowledged influence on Pen-guin was Albatross Books, founded in Hamburg in 1932, which not only produced thoughtfully designed literary works as mass-market paperbacks but also colour-coded its titles according to genre, the way Penguin would later do.

And as one of Lane's executives admitted, Penguin's avian name may have 'subconsciously hatched from an Alba-tross egg' too. The hatching took place when, at a meeting to discuss what the new imprint should be called, some-one who wasn't officially there made a world-changing suggestion. Overhearing the discussion, Joan Coles, a sec-retary, paused from typing on the other side of a partition to say, 'What about Penguins?' Lane approved at once of the name's 'dignified flippancy' and sent his designer Edward Young off to London Zoo to sketch a few of the real things as part of what he described as 'the first serious attempt at introducing "branded goods" to the book trade'.

So it was that on 30 July 1935 the first ten Penguin books appeared in covers featuring a flightless bird and those soon-to-be-familiar horizontal stripes: orange for fiction, blue for non-fiction, green for crime.[9] Now all Lane had to do was sell them . . .

For a while, it looked as if this might be a problem serious enough to scupper the whole plan. Lane had printed 20,000 copies of each title and needed to sell between 17,000 and 18,000 merely to break even. Many booksellers refused to take any, including – disastrously – W. H. Smith. 'It's a flop,' Lane told a friend. 'I've got to pack it up.' But then, three weeks before publication, he called on Clifford Prescott, Woolworth's head buyer for books (and haberdashery). Prescott wasn't keen either – until his wife showed up for a lunch date and told him how good she thought the books were. A couple of days later Woolworth's ordered 63,500, influencing other shops to take them too, and with that Penguin was made. It went on to sell a million books in its first four months and three million in its first year – by which time it had been incorporated as a company of its own, independent of The Bodley Head.[10]

When it struck out on its own, Penguin was initially

---

9   Nowadays orange is also used for quiz books.
10 These days, in a full-circle kind of way, The Bodley Head is part of Penguin Random House, where it specialises in serious non-fiction.

based in Great Portland Street in Central London, with its warehouse in the mouse-infested crypt of Holy Trinity Church, Marylebone, where the toilet facilities consisted of a single tin bucket. By 1937 the company was doing well enough to build its own office-warehouse combo on a cabbage field in Harmondsworth, Middlesex, bought from a farmer for a little over £2,000 – plus £200 for the cabbages.

Lane's faith in the public's intelligence was further vindicated when Pelican was launched and made bestsellers out of works on economics, psychiatry, art and ancient history. The lofty aim was to bring what Lane called 'the finest products of modern thought and art to the people' – and, according to the *Spectator* in 1938, Pelicans became 'a fact of enormous importance in the struggle to overcome economic restrictions to knowledge'. As war approached, Lane expanded his company's non-fiction into more topical books, known as Penguin Specials, which were published at a speed that modern publishers might find hard to believe, let alone match. At the time of the 1938 Munich Crisis, Shiela Grant Duff's *Europe and the Czechs* was in the shops less than a week after she delivered the typescript.

Once the conflict started, other publishers soon had another reason to resent Penguin's huge pre-war success. Paper rationing was based on how much paper each publisher had used between August 1938 and August 1939 – and because Penguin had been so productive that year it now had a lot more of the stuff than its rivals. Over the next six years, Penguin published more than 600 books, among them such bestsellers as *Aircraft Recognition* and *Keeping Poultry and Rabbits on Scraps*. With many male staff away at war themselves, much of the work was done by Eunice

Frost, one of Britain's first women editors. Frost had joined Penguin as Lane's secretary in 1937, but shortly afterwards he asked her, 'How do you like reading?', pushed a pile of books across the desk and gave her an editorial role.

The war years also saw an enormous surge in reading more generally, with civilians using books to get them through the boredom of the blackouts and people in the military needing to kill time on long journeys and in barracks. However, in a possible foreshadowing of the company's problems in the 1950s, Penguin's Forces Book Club (FBC) was one of its rare early failures. The soldiers to whom the specially selected titles were sent would, it transpired, have preferred Westerns to the well-intentioned likes of *Growing Up in New Guinea* by Margaret Mead – and many FBC Penguins ended up doing their bit for troop morale as fuel and toilet paper.

Even so, Penguin entered the post-war world on a confident high: not surprisingly given that it was a world the company had done much to create. Once victory seemed likely, its non-fiction books were at the centre of the debate as to what sort of Britain should emerge from the conflict, and on the whole they argued for precisely the kind of welfare state that the Labour government built. Penguin now took its place alongside the BBC and the NHS as one of the benign virtual-monopolies that dominated British life: 'not so much a publisher as an estate of the realm,' as the writer and all-round man of letters John Gross later put it. Its stock rose higher still with the introduction of Penguin Classics – in those days translated works only – which turned Homer, Cervantes and Dante into three of Britain's bestselling authors.

To polish his brand even further, Lane took advice

on where to find the best book designer in Europe. He was given the name of Jan Tschichold, who as a young man in Germany had been arrested by the Nazis for the unusual offence of criminal typography.[11] By 1947 Tschichold was living in Switzerland and once Lane brought him to Penguin, he overhauled and professionalised every aspect of production to make the books both more elegant and more uniform. He also slimmed down the penguin of the logo.

Meanwhile back in the world of bees (remember them?), quizzes had had a pretty good war too. They'd continued as a radio staple, but had also diversified into the type of live events that today's pub quizzers would surely recognise – even if they mightn't have welcomed the questions set by Women's Institutes and Mothers' Unions on road safety and farming. And yet, as it turned out, both quizzes and Penguin had a tricky decade ahead, and for fundamentally the same reason . . .

The 1950s are not often seen as a flashy decade, but the growth of consumerism meant that both Penguin and the BBC were faced with new, brasher, American-influenced competitors, who transformed paperback publishing and the British quiz.

For Penguin, those competitors were the likes of Pan, Corgi and Fontana, with their vulgar picture covers and their even more vulgar sales figures. ('Breastsellers' was Lane's usual description of their books.) Suddenly, Penguin's understatedly stylish designs began to seem a bit old-fashioned, causing the company to experiment rather

---

11 Apparently his fondness for slanted lettering made his work 'Bolshevik'.

half-heartedly with picture covers of its own. The results, as Jeremy Lewis puts it in his highly recommended *Penguin Special: The Life and Times of Allen Lane* (published, as luck would have it, by Penguin), 'ended up by pleasing no one, being too tasteful for the vulgarians and too radical for the old guard'.

For the BBC, the brasher competitor was, of course, ITV, which started in 1955 and – like those new paperback publishers – wasn't afraid to give the punters what they wanted, rather than what they should want. (For the BBC's first director-general Lord Reith, the ideal was giving the public something 'slightly better than it now thinks it likes'.) And what the punters wanted, it seemed, was not just quizzes but big-money quizzes. True, successful competitors in the BBC's *Have a Go* radio show could bag themselves £1 18s 6d. But in the week it launched, ITV brought the nation *Double Your Money*, presented by Hughie Green, which offered a top prize of £1,000.

Naturally, not everybody approved of the blizzard of similar 'give-away' programmes that ITV was soon serving up. The cultural critic Richard Hoggart – who also acted as an unpaid adviser to Penguin – stated regally, 'We don't like the quiz shows. They pander to the need for quick money.' The trouble was that, as with Pan, Corgi and Fontana, the public appeared to like them quite a lot.

In the late 1950s, American TV's big-money quiz shows found themselves engulfed in scandal when it emerged that they were giving their favoured competitors the answers in advance. Not only that, but the producers also supplied acting instructions. 'I was told,' remembered a contestant on the show *Dotto*, 'how to bite my lips, clench my fists and look agonized as I supposedly struggled to find the answers. They even told me how, at the last moment, to make my face light up as if the answer had suddenly come to me.' The whistle was finally blown by Herb Stempel, a Bronx Jew, who'd been ordered to lose to Charles Van Doren, a university professor from an impeccably Wasp family. Stempel's breaking point came when he was instructed to answer 'Which motion picture won the Academy Award in 1955?' with 'On the Waterfront' – when, as he well knew, the real answer was Marty, set in the Bronx and one of his favourite films.

While there was no suggestion that British shows were doing the same thing – although some did supply helpful reading lists – the American scandals meant they now faced increasing scrutiny from the sort of Hoggartians who'd despised them all along. In 1960 the UK government set up the Pilkington Committee on the future of broadcasting. Its report, published two years later, took a particularly dim view of big-money quizzes, with their 'appeal to greed and fear'. It also recommended that 'the maximum value of prizes should be greatly reduced'.

Three months later, in an obvious attempt to demonstrate that it was moving quizzes upmarket, ITV introduced University Challenge – 'a quiz,' in the words of presenter Bamber Gascoigne, 'that the public wouldn't know the answers to'. Even so, by the late 1960s the show was pulling in ten million viewers a week, meaning that Bill Wright, the new head of the BBC's Outside Broadcast Quiz Unit, was keen to hit back with a brainy quiz of his own. His eureka moment came when he decided to draw on his experience of being interrogated by the Gestapo during the war – and the result was Mastermind. Instead of demanding name, rank and number, presenter Magnus Magnusson asked for name, occupation and specialised subject, but the voice of the interrogator still came out of darkness to someone sitting in a brightly lit chair.

As for the sort of quizzes the Pilkington Report had attacked, the prize money wasn't reduced – but nor was it allowed to rise too much. For the next 30 years, £6,000 was the maximum amount that anybody could win on UK television, as the British TV quiz entered what Alan Connor calls 'a three-decade holding pattern'.

For Penguin, the 1960s began more promisingly, when it published the unexpurgated version of D. H. Lawrence's *Lady Chatterley's Lover*, was prosecuted for obscenity at the Old Bailey and won.

The prosecution was led by Mervyn Griffith-Jones QC, who possibly supplied too much information when he explained beforehand: 'I put my feet up on the desk and start reading. If I get an erection, we prosecute.' His most famous – and ill-advised – moment in the trial itself was when he asked the jury if *Lady Chatterley* was a book you would 'wish your wife or servants to read'. For its part, the defence called several eminent figures to testify to the novel's literary worth, including Richard Hoggart and the Bishop of Woolwich, who caused a press sensation with his claim that every Christian should read it. Sadly, Enid Blyton turned down Penguin's invitation to testify too. Understandably astonished to be asked, she replied that she'd never read the novel and 'my husband said NO at once'.

After Penguin was acquitted, *Lady Chatterley's Lover* duly became its bestselling book ever, with the paperback bearing a dedication 'to the twelve jurors'. The following year, Penguin was floated as a public company that, awash with *Chatterley* cash, proved predictably attractive to investors and Lane became a millionaire. But the challenges posed by those rival publishers hadn't gone away – and the question of how much Penguin should change its trusty methods would prove increasingly divisive as the decade went on.

In 1966 Lane acknowledged, 'I find myself out of sympathy with much of contemporary literature, theatre, typography and art.' Nonetheless, he did allow Penguin covers to be redesigned in keeping with the general 1960s trend towards more colour and glossiness. E. V. Rieu, who'd edited Penguin Classics since they began, was aghast at the colour reproductions of paintings that now adorned his books. ('Oh my poor series, of which I used to be so proud,' he lamented.) Anthony Powell was so affronted that he went off to Fontana.

But as the debate/ructions between the old guard and the company's handful of young turks continued, Penguin did have one unalloyed triumph throughout the 1960s. When it started in 1940, the Puffin imprint for children consisted of non-fiction picture books, with titles like *The Wonders of Sea Life* and, less alluringly, *The Story of Furniture*. The first batch of fiction – led by Barbara Euphan Todd's *Worzel Gummidge* – came in 1941, and over the next 20 years Puffin's hits included E. B. White's *Charlotte's Web*, C. S. Lewis's *The Lion, the Witch and the Wardrobe* and Norman Hunter's *Professor Branestawm* series (to mention just three that won't give away any answers to the quiz questions in this book).

The creator of Puffin, and the person responsible for all these successes, was Eleanor Graham – which makes it a bit unfair that the name that's always associated with the imprint is Kaye Webb. Still, there's no denying that, after taking over in 1961, Webb's achievements were staggeringly impressive. Making the most of a boom in children's books, for which she was, of course, partly responsible, Webb increased the number of Puffin books from 150 to around 800 by the end of the decade. In 1967 she founded the fondly remembered Puffin Club, which had 44,000 members within two years.

Puffin also came to the rescue in the 1970s when Penguin's adult publishing was in something of a crisis. Books that it would once have snapped up were now coming out from the newly established paperback divisions of their original publishers. Meanwhile, the death of Allen Lane in 1970 was perhaps the literary equivalent of Alex Ferguson's retirement as Manchester United's manager – when a man who'd been so dominant for so long proved not just a hard act to follow but, very probably, an impossible one.

Happily, the 1980s saw both a revival in fortunes for Penguin and a great leap forward for quizzing – and, it being the 1980s, money and big business go a long way to explaining why.

The turning point for quizzing was the brainchild of two Canadian hippies. They were Chris Haney and Scott Abbott, who in 1980 decamped to Spain with a set of reference books, enough money to keep themselves in beer and a peculiar idea: that quizzing – which in America still hadn't recovered from the scandals of the 1950s – could be the basis not of a TV show where people won cash but a board game for which they'd *pay* cash.

Now that Trivial Pursuit has become part of social history, it's easy to forget just how peculiar this idea seemed – not least to America's leading board-game companies, MB and Parker Brothers, both of whom turned it down. In the end, Haney and Abbott were forced to manufacture it themselves, before licensing it to Selchow & Richter. They then received a game-changing boost when Glenn Close revealed to *Time* magazine that she and her fellow actors in the 1983 baby-boomer film *The Big Chill* had become 'addicted to' Trivial Pursuit, 'playing it night and day'.

In 1984 the new game shifted 22 million units in America alone, and was poised to take its place at dinner parties (often held by what were then known as yuppies) all over the world. But first it had to go through a significant court case of its own.

Fred L. Worth was an air traffic controller by day and an assiduous collector of pop culture trivia by night. In 1974 he published a large collection of his most treasured facts in *The Trivia Encyclopaedia*, which like many reference books took the precaution of throwing in a deliberately false piece of information so that he'd know if anybody plagiarised his work.[12] The first name of the television detective Columbo, he told his readers, was Philip – even though the writers deliberately didn't give him a first name.

Ten years later, he picked up Trivial Pursuit and thought, 'Boy, this stuff sure looks familiar.' And, indeed, it sure would have done. The game had lifted over a quarter of its questions from his work, complete with the same occasional mistakes and misprints. Most significantly of all, the answer to the Entertainment question 'What is Columbo's first name?' was 'Philip'. And so, in October 1984, Worth sued Haney and Abbott for $300 million, mortgaging his house to pay the legal fees.

---

12 The *London A–Z*, for instance, includes around 100 streets that don't exist in real life so as to catch out other firms trying to pass off its maps as their own.

To his horror, however, he lost the case. Haney and Abbott didn't deny that they'd turned more than a thousand of his facts into questions, mainly by the cunning method of rearranging them slightly and adding a question mark. Instead, their defence was that the whole point of encyclopaedias and reference books is to provide information for other people to use in any way they like. The court backed this belief, ruling that the 'discovery of a fact, regardless of the quantum of labor and expense, is simply not the work of an author' – in short, that facts can't be copyrighted.

Columbo's first name wasn't the only mistake in Trivial Pursuit. Haney and Abbott even fell for a spoof history of the bra by the humorist Wallace Reyburn – which is why their answer to 'Who invented the brassiere?' was 'Otto Titzling'. Nevertheless, the game's huge success led in turn to the revival of quiz shows on television – culminating in *Who Wants to Be a Millionaire?*, which started on ITV in 1998 and has since spread to around 160 countries. It also led in Britain to the unstoppable rise of the pub quiz, the charity quiz, the school quiz and the literary festival quiz. Less unstoppable, mind you, was the rise of the quiz machine, which became such a reliable source of income for serious quizzers (some earning £60,000 a year from them) that the makers were obliged to add the kind of impossible questions that put ordinary players off.

Essentially, though, quizzing hasn't really looked back since Glenn Close gave that *Time* interview. These days, a batch of general knowledge questions is as much a staple of newspapers as the crossword, smartphones have scores of quizzing apps available and, should you wish, you can spend hours online answering questions that will prove once and for all which Disney princess you most resemble or which Ariana Grande song most accurately describes your life.[13] According to a poll commissioned by Alan Connor, 81 per cent of British adults watch, listen to or take part in quizzes, and 44 per cent do so every week. (And, fortunately, I don't think this includes the Disney/ Ariana sort.)

On TV, too, quizzes are now more widespread than ever and include plenty of shows that confound all theories about Britain dumbing down. Despite its fiendishness, *Only Connect* – where teams have to find the punishingly difficult links between punishingly difficult things – is a reliable ratings hit for BBC2. And if you're one of those people who like to think that *University Challenge* has got easier over the years, I'd suggest a quick visit to YouTube, where Bamber Gascoigne can be seen asking the students of the 1960s questions that the average pub quizzer of today would regard with scorn. ('Which god is Wednesday named after?')

Where Penguin is concerned, the money and big-business side of the story is more complicated, comprising – as in the wider publishing industry – a lengthy series of take-overs and mergers. In fact, these began the day after Allen Lane died, when Penguin merged with Pearson Longman. But only in the 1980s, under its new American CEO Peter Mayer, did Penguin begin its transition into a full-on global brand.

When Mayer was appointed in 1978, Penguin was, by common agreement, somewhat moribund. He himself found it 'toffee-nosed' about commercial fiction, while

---

13 In my case, Tiana and 'thank u, next' respectively.

the continuing growth of 'vertical publishing' (the same publisher releasing the hardback and paperback editions) meant that it was in danger of losing – or had already lost – the type of books it had always relied on. Its reputation as a cultural institution may still have been secure, but its future wasn't.

Mayer's solution was to combine Penguin's longstanding – and his own – erudition with a far more businesslike approach. As part of a move towards the more aggressive marketing of commercial titles, the company threw its weight behind such distinctly non-Lane-like books as Shirley Conran's *Lace*: the novel that contains the celebrated line 'Which one of you bitches is my mother?'[14] To ensure a healthy supply of paperbacks, Mayer brought three new hardback imprints to Penguin in Britain: acquiring Michael Joseph and Hamish Hamilton, and importing Viking, which the company already owned in America. He also bought up Beatrix Potter's publisher Frederick Warne and oversaw the expansion of Penguin in the USA, Canada, Australia, New Zealand, India and South Africa.

But, in more disappointing news for dumbing-down theorists, Mayer's unashamed commerce didn't prevent him from retaining a commitment to serious literature – most notably in 1989 when, after Ayatollah Khomeini's fatwa, he refused to withdraw Salman Rushdie's *The Satanic Verses*. 'Once you say I won't publish a book because someone doesn't like it or someone threatens you, you're finished,' he said. 'Some other group will do the same thing, or the same group will do it more.' In something of a return to first principles, he celebrated the company's sixtieth

---

14 In 1984 *Lace* became a miniseries – and in 1993 readers of America's *TV Guide* voted that line the greatest ever in television history.

anniversary in 1995 with the Penguin 60s: mini-books extracted almost exclusively from the highbrow end of Penguin's back catalogue, which cost 60p each and between them sold 30 million copies.

Mayer retired in 1997, but – fortunately for my purposes, given that all my questions are about books published by Penguin and its sister publishers – the mergers and take-overs continued. The big one came in 2013, when Penguin and all its affiliates joined Random House and all its affili-ates to become, as you might expect, Penguin Random House, an international corporation that's now the largest publisher in Britain – and the world. Penguin's influence is felt, too, in the scores of films and TV programmes that its books have inspired – which is why, while most of the questions here are book-related, there's a fair smattering designed to test your knowledge of other media.

Penguin has, in other words, come a long way since Edward Young sketched those birds in London Zoo . . .

But in case this conclusion seems too triumphalist about both Penguin and quizzes, there's maybe one last thing that they've shared in more recent years: a fear of what the new digital world might mean. A celebratory corpor-ate history of Penguin produced in 2009 couldn't disguise its deep anxiety about the possible disappearance of the physical book – although when contemplating an utterly changed world it prefers the customary euphemism 'excit-ing' to, say, 'terrifying'. (What it didn't foresee was the rise of audiobooks, which are now a much-loved part of many people's reading lives.) In *The Joy of Quiz*, Alan Con-nor quotes Larry Page, the co-founder of Google, saying, 'Eventually you'll have the implant, where if you think about a fact, it will just tell you the answer.' So, Connor wonders, might all quizmasters one day have to begin with 'an instruction for all contestants to deactivate their implants'?

Without wishing to sound unimaginative, my own answer to this question would be 'no' – just as, ten years on, that corporate nervousness about the end of the printed book already sounds faintly old-fashioned, now that the e-book has failed to destroy the world of publishing any more than radio and cinema did. Or before that, newspapers. Or after that, television.

And in that spirit, let me offer you this book – confident that at this very moment many of you will be holding something made of paper, and that none of you will have to deactivate your implants.

HOW IT WORKS:
GUIDELINES TO THE
QUIZZES

The main guideline for the quizzes that follow is that you should, of course, feel free to take no notice of my guidelines and use the book any way you like. However, for those people (perfectly normal in my opinion) who want to turn it into a competition, here's my suggested scoring system.

In the quickfire rounds that bookend each of the ten quizzes, how about two points for a perfect answer and one if you're nearly there? (Personally I'm never a fan of the half-point.) As a bonus, the title of each of these rounds is also the title of a Penguin Random House book – some harder than others – and there's two more points if you can identify who wrote it.[1] Alternatively, although it may stick in the craw of some quizzers, you could leave the competition behind for the bonus and do that bit together.

Round 2 in every quiz is Name the Author – where there are four clues to a writer's identity in order of what's meant

---

[1] And well done if you spotted that the main title of this section refers to the bestselling 'How It Works' series of Ladybird books for grown-ups by Jason Hazeley and Joel Morris.

to be decreasing difficulty. My suggestion here is pretty much as you'd imagine: four points (and quite a lot of glory) if you get the answer after one clue, three points (and only slightly less glory) if you get it after two, and so on.

Next comes an Extracts round, where the scoring system varies more than in the other rounds. As a result, I've explained how it might work at the start of each one.

Round 4 is a more cryptic affair that comes in three forms over the course of the book: Connect Three, Odd One Out and Order, Order. Of all the rounds, this is the one that might lead to most discussion/argument among those playing for points. But I'd propose a maximum of five if you get the basic answer *and* supply all the details as to why – i.e. name all the books and authors that the clues refer to – with fewer from there depending on how much of the full explanation is missing. Admittedly, in some of the questions, there aren't so many details to fill in, but let's not get too fiddly here. Like all the other guidelines, this one is for you to use, adapt or completely ignore as you see fit. After all, was it not the great eighteenth-century critic Joseph Addison who said, 'There is sometimes a greater judgement shown in deviating from the rules . . . than in adhering to them'?[2]

Happy quizzing!

---

2 Yes, it was.

QUIZ I

ROUND I

QUICKFIRE:
THE FOUNTAINHEAD

All the answers here begin with A . . .

I.

Who's the main human character in Douglas Adams's *The Hitchhiker's Guide to the Galaxy*?

2.

Lestat de Lioncourt is the main character in whose novel series *The Vampire Chronicles* – which began in 1976 with *Interview with the Vampire*?

3.

The speech beginning 'All the world's a stage' is in which Shakespeare play?

3

ANSWERS ON PP. 21–3

4.

Which chef's books include *Ultimate Barbecue Bible*, *Caribbean Kitchen* and *The Top 100 Recipes from Ready Steady Cook!*?

5.

Who wrote the world's best ever selling book by a teenager, *The Diary of a Young Girl*?

6.

Which philosopher was a student of Plato and a tutor of Alexander the Great?

7.

Who has published several volumes of his diaries with subtitles that include *Prelude to Power 1994–1997* and *Countdown to Iraq*?

8.

Who preceded Carol Ann Duffy as the Poet Laureate?

9.

What's the title of the classic First World War novel by Erich Maria Remarque?

10.

Who's the creator of the modern detective Daniel Hawthorne, the wartime detective Christopher Foyle and the teenage spy Alex Rider?

11.

Oliver Bowden has written several books based on which video game that features the Knights Templar as the main baddies?

12.

Whose autobiography did Gertrude Stein write in 1933?

5

ANSWERS ON PP. 21–3

QI

ROUND 2

NAME THE AUTHOR

*Can you guess the writer from these clues
(and, of course, the fewer you need the better)?*

I.

A.   In later life, she was a prize-winning breeder of sheep.

B.   Despite being British, she was played by Renée Zellweger in a 2006 biopic.

C.   Her former home, Hill Top Farm in Cumbria, has been open to the public since 1946.

D.   The main character in her first book had siblings called Flopsy, Mopsy and Cotton-tail.

6

2.

A. He's the only writer to have a British football club named after one of his novels.

B. The biggest monument to any novelist anywhere in the world is to him.

C. Both the football club and the monument are in Edinburgh.

D. He's generally credited with establishing the form of the historical novel as we know it today – through novels such as *Ivanhoe* and *Rob Roy*.

3.

A. His brother Gregory is a composer, whose works include *Missa Charles Darwin*.

B. He wrote the world's bestselling novel of the first decade of the twenty-first century.

C. Its main character is Robert Langdon . . .

D. . . . and it has the name of an artist in the title.

7

QI

4.

A. In her time as a literary critic, she wrote an essay called 'Silly Novels by Lady Novelists'.

B. She caused something of a scandal by not being married to the man she lived with for more than 20 years, until his death in 1878.

C. Her real name was Mary Ann Evans.

D. Both Martin Amis and Julian Barnes have described her 1872 book that features the characters Dorothea Brooke, Edward Casaubon and Dr Tertius Lydgate as the greatest English novel of them all.

5.

A. His wife's maiden name was Sayre.

B. He was named after a distant cousin who wrote the lyrics for what later became the American national anthem.

C. He coined the phrase 'the Jazz Age', for the 1920s, in the title of a short-story collection.

D. His last completed novel was *Tender Is the Night*.

ANSWERS ON PP. 24–6

6.

A. Her famous mother died from puerperal (childbed) fever ten days after giving birth to her.

B. That famous mother wrote *A Vindication of the Rights of Woman* in 1792.

C. Her best-known novel was published when she was 20, and has been the subject of nearly 100 film adaptations, some of them quite loose . . .

D. . . . and it features a monster that's never named.

9

ROUND 3

EXTRACTS:
HOW IT ALL BEGAN

The following passages are the first words of the first novels their authors ever published. Can you name the author in each case for one point, and the novel for another?

I.

From 1987:

—We'll ask Jimmy, said Outspan.—Jimmy'll know.

Jimmy Rabbitte knew his music. He knew his stuff alright. You'd never see Jimmy coming home from town without a new album or a 12-inch or at least a 7-inch single. Jimmy ate Melody Maker and the NME every week and Hot Press every two weeks. He listened to Dave Fanning and John Peel. He even read his sisters' Jackie when there was no one looking. So Jimmy knew his stuff.

2.

From 1992:

The snow in the mountains was melting and Bunny had been dead for several weeks before we came to understand the gravity of our situation. He'd been dead for ten days before they found him, you know. It was one of the biggest manhunts in Vermont history – state troopers, the FBI, even an army helicopter; the college closed, the dye factory in Hampden shut down, people coming from New Hampshire, upstate New York, as far away as Boston.

3.

From 1989:

The old woman remembered a swan she had bought many years ago in Shanghai for a foolish sum. This bird, boasted the market vendor, was once a duck that stretched its neck in hopes of becoming a goose, and now look! – it is too beautiful to eat.

Then the woman and the swan sailed across an ocean many thousands of li wide, stretching their necks toward America. On her journey she cooed to the swan: 'In America I will have a daughter just like me. But over there nobody will say her worth is measured by the loudness of her husband's belch. Over there nobody will look down on her, because I will make her speak only perfect American English.'

ANSWERS ON PP. 27–8

4.

From 2000:

Early in the morning, late in the century, Cricklewood Broadway. At 06.27 hours on 1 January 1975, Alfred Archibald Jones was dressed in corduroy and sat in a fume-filled Cavalier Musketeer Estate face down on the steering wheel, hoping the judgement would not be too heavy upon him. He lay forward in a prostrate cross, jaw slack, arms splayed either side like some fallen angel; scrunched up in each fist he held his army service medals (left) and his marriage licence (right), for he had decided to take his mistakes with him. A little green light flashed in his eye, signalling a right turn he had resolved never to make. He was resigned to it. He was prepared for it. He had flipped a coin and stood staunchly by its conclusions. This was a decided-upon suicide. In fact it was a New Year's resolution.

ANSWERS ON PP. 27–8

ROUND 4

CONNECT THREE

Can you link the three items in each case?

I.

The surname of Michael, John and Wendy in *Peter Pan*

What Winnie-the-Pooh goes in quest of in the first chapter of *Winnie-the-Pooh*

What a Walter Greenwood novel had on the dole, and a Gabriel García Márquez novel had in the time of cholera

2.

The surname shared by Jared, the author of 1997's Pulitzer Prize-winning *Guns, Germs and Steel*, and John, Nigella Lawson's late husband, who chronicled his final illness in *C: Because Cowards Get Cancer Too*

13

ROUND 4

Carson McCullers's lonely hunter in a novel of 1940

The detective created by Dashiell Hammett who first appeared in *The Maltese Falcon* and was played on screen by Humphrey Bogart

3.

The author of *The Razor's Edge* and *Of Human Bondage*

Regan's husband in Shakespeare's *King Lear*

An 1896 poetry collection by A. E. Housman

4.

M. M. ____, author of the 1,000-page novel about British India *The Far Pavilions*, which became HBO's first drama serial

A Dark-Adapted ____, a psychological thriller by Barbara Vine

The Cruel ____, a Second World War novel by Nicholas Monsarrat

ANSWERS ON PP. 29–30

5.

The member of the royal family to whom Jane Austen
dedicated *Emma*

The city where the heroine of Philip Pullman's *His Dark
Materials* grows up – albeit in a parallel universe

The main character of Sebastian Faulks's *Devil May Care*
and William Boyd's *Solo*

6.

Henry James's Washington and Patrick Hamilton's Hangover

The Bertolt Brecht play whose German title is *Der kauka-
sische Kreidekreis*

The film series that has inspired dozens of spin-off novels,
including *Tatooine Ghost*, *The Cestus Deception* and *Lords of
the Sith*

ANSWERS ON PP. 29–30

ROUND 5

QUICKFIRE:
WAYS OF SEEING

Twelve film and TV questions . . .

1.

Who's the author of the novels *Playing with Fire, Make My Wish Come True* and *He's the One* – and met her first husband when they both appeared in the third series of *I'm a Celebrity . . . Get Me Out of Here!*?

2.

Who created the teenage criminal mastermind Artemis Fowl, as featured in a 2019 film directed by Kenneth Branagh?

3.

Who wrote the 1994 novel *What a Carve Up!*, which took its title from a British comedy-horror film of 1961?

16

ANSWERS ON PP. 31–3

4.

Who created the 1978 children's book that has no words and was turned into an animated film that has no words either – apart from a song sung by the choirboy Peter Auty?

5.

Daniel Radcliffe made his first post-Harry Potter film appearance in which horror movie, based on a novel by Susan Hill?

6.

Push, the 1996 debut novel by Sapphire about an obese Harlem teenager, was turned into which film – for which Geoffrey S. Fletcher became the first African-American to win an Oscar for screenplay-writing?

7.

Who played one of the main characters in the 2005 TV version of her own novel *Life Isn't All Ha Ha Hee Hee*?

8.

The 2016 film *Me Before You* is based on a bestselling novel by whom?

QI

9.

Which film that regularly appears in the top ten in critics' polls of the greatest ever movies was inspired by Arthur C. Clarke's short story 'The Sentinel'?

10.

Which writer's works have been adapted into the films *Minority Report*, *Total Recall*, *Blade Runner* and the Amazon TV series *The Man in the High Castle*?

11.

Two of Damon Runyon's stories about Prohibition-era New York were turned into which musical – later a film starring Frank Sinatra and Marlon Brando?

12.

Kim Novak, Julia Foster, Robin Wright and Alex Kingston have all played which Daniel Defoe character on screen?

ANSWERS
TO QUIZ I

ROUND 1

1.

Arthur Dent

2.

Anne Rice

3.

As You Like It. It's the speech that goes on to describe the seven ages of man from 'the infant/ Mewling and puking in the nurse's arms' to the old man 'Sans teeth, sans eyes, sans taste, sans everything'.

4.

Ainsley Harriott

5.

Anne Frank

6.

Aristotle

7.

Alastair Campbell

8.

Andrew Motion

9.

All Quiet on the Western Front

10.

Anthony Horowitz

11.

Assassin's Creed

12.

Alice B. Toklas. *The Autobiography of Alice B. Toklas* is a memoir of Stein's years as a salon host in Paris written in the voice of her life partner.

Bonus:

The Fountainhead is a 1943 novel – and the first major success – by Ayn Rand. A hymn in praise of individualism, the book is an acknowledged influence on American conservatives, including Donald Trump, who once said: 'It relates to business [and] beauty [and] life and inner emotions. That book relates to . . . everything.' He also admits to identifying with its anti-establishment hero, Howard Roark.

ROUND 2

I.

Beatrix Potter, whose first book was *The Tale of Peter Rabbit*. The film with Renée Zellweger was *Miss Potter*, directed by Chris Noonan, who admitted he hadn't read Beatrix Potter growing up in Australia but was 'aware of her because of all that crockery with her characters on it'.

2.

Walter Scott. The football club is Heart of Midlothian – Hearts for short. (Edinburgh's main railway station, Waverley, is also named after a Scott novel.) Before he turned to fiction in 1814, Scott was an enormously successful poet who the previous year had declined the post of Poet Laureate, because he felt the holders of the office had become 'a succession of poetasters [who] had churned out conventional and obsequious odes on royal occasions'. He quickly became so popular as a novelist that when *Rob Roy* was published in 1817 a ship sailed from Leith to London filled with nothing but copies of the book.

3.

Dan Brown, author of *The Da Vinci Code*, the second in the Robert Langdon series (following *Angels & Demons* – which made little impact when it was first published but duly became a bestseller itself after *The Da Vinci Code* hit the big time). *Missa Charles Darwin* mixes the traditional structure of a mass setting with Charles Darwin texts. It also has a chapter dedicated to it in Brown's 2017 novel *Origin*.

4.

George Eliot, whose much-admired 1872 novel is *Middle-march*. Her live-in lover was the philosopher and critic George Henry Lewes, whose wife already had two children by another man. He couldn't, however, divorce her for adultery because he'd agreed to it under the terms of their open marriage. (We're clearly at the racier end of the Victorian spectrum here.) Two years after he died, George Eliot married a man 20 years younger than her.

5.

F. Scott Fitzgerald, where the F stands for Francis – as in Francis Scott Key, whose poem 'Defence of Fort M'Henry' became 'The Star-Spangled Banner'. The collection *Tales from the Jazz Age*, published in 1922, includes the story 'The Curious Case of Benjamin Button', which was made into the 2008 film starring Brad Pitt as Benjamin, who ages in

AI

reverse. Fitzgerald's marriage to Zelda (née Sayre) was famously tempestuous, what with him being a drunk and her suffering from mental illness – both of which are reflected in the portrayal of Dick and Nicole Diver, the main characters in *Tender Is the Night*.

6.

Mary Shelley, the daughter of Mary Wollstonecraft and the author of *Frankenstein* (named, of course, after the doctor who creates the monster). As for those film adaptations, some of the looser ones are *Jesse James Meets Frankenstein's Daughter*, *Frankenstein Meets the Space Monster* and *Alvin and the Chipmunks Meet Frankenstein*.

ROUND 3

1.

Roddy Doyle's *The Commitments*, the tale of a Dublin soul band that's since become a hit film and a hit stage musical. (Younger quizzers might have to ask their parents about *Melody Maker*, John Peel and *Jackie* magazine – although Dublin's *Hot Press* and radio DJ Dave Fanning should be familiar to anybody from Ireland.)

2.

Donna Tartt's *The Secret History*, in which a group of classics students at the fictional elite college of Hampden in Vermont murder one of their classmates. (No spoiler alert needed here, as the very next paragraph makes it clear who's responsible for Bunny's death.)

3.

Amy Tan's *The Joy Luck Club*, a novel about Chinese immigrant mothers and their Chinese-American daughters in San Francisco. The club of the title is centred mainly around food and mah-jong.

4.

Zadie Smith's *White Teeth*. Luckily, Archie is parked in the delivery area of a local halal butcher who needs the area kept free – and pulls the car window down on the grounds that 'we're not licensed for suicides around here. This place halal ... If you're going to die round here, my friend, I'm afraid you've got to be thoroughly bled first.'

ROUND 4

1.

Terms of endearment: Michael, John and Wendy's surname is <u>Darling</u>; Winnie-the-Pooh characteristically goes in quest of <u>honey</u>; Greenwood wrote <u>Love</u> on the Dole, a classic of working-class northern life in the 1930s – and García Márquez, <u>Love</u> in the Time of Cholera.

2.

Suits of playing cards: Jared and John <u>Diamond</u>; McCullers's The <u>Heart</u> Is a Lonely Hunter; Hammett's Sam <u>Spade</u>. (Guns, Germs and Steel, incidentally, has the ambitious but accurate subtitle A Short History of Everybody for the Last 13,000 Years.)

3.

British counties: <u>Somerset</u> Maugham wrote The Razor's Edge and Of Human Bondage; Regan married the Duke of <u>Cornwall</u> (the one who plucks out Gloucester's eyes); and that Housman poetry collection was A <u>Shropshire</u> Lad.

4.

Words that sound like individual letters: M. M. <u>Kaye</u> wrote *The Far Pavilions*; *A Dark-Adapted <u>Eye</u>* was the first novel that Ruth Rendell wrote under her recurring pen name of Barbara Vine; *The Cruel <u>Sea</u>* is Monsarrat's tale of the Royal Navy during the Battle of the Atlantic, based on his own experiences, which became a much-loved British war movie.

5.

The green set of streets on a traditional Monopoly board: *Emma* is dedicated to the Prince <u>Regent</u>; Pullman's setting is <u>Oxford</u>; Boyd's and Faulks's novels both star James <u>Bond</u> – in two of the sequels written by authors other than Ian Fleming.

6.

Shapes: James's *Washington <u>Square</u>* is a novel from 1880, and Hamilton's *Hangover <u>Square</u>*, his tale of rackety London drinkers in the 1930s, was published in 1941; Brecht's play in English is *The Caucasian Chalk <u>Circle</u>*; and the film series is *<u>Star</u> Wars* – where Tatooine and Cestus are planets and the Sith are sworn enemies of the Jedi.

ROUND 5

1.

Katie Price (aka Jordan) – that first husband being Peter André, whom she married in 2005. The couple separated in 2009.

2.

Eoin Colfer

3.

Jonathan Coe

4.

Raymond Briggs – the book was *The Snowman*. (The Aled Jones version of 'Walking in the Air' was first recorded a couple of years later for a Toys "R" Us advert.)

5.

The Woman in Black

6.

Precious. (The film's original title was *Push: Based on the Novel by Sapphire*.)

7.

Meera Syal

8.

JoJo Moyes

9.

2001: A Space Odyssey. (Clarke co-wrote the screenplay with the director, Stanley Kubrick.)

10.

Philip K. Dick

11.

Guys and Dolls. (The two stories are 'The Idyll of Miss Sarah Brown' and 'Blood Pressure'.)

12.

Moll Flanders – from the novel of the same name, although technically the book's full title is (possible spoiler alert): *The Fortunes and Misfortunes of the Famous Moll Flanders, Who was born in Newgate, and during a Life of continu'd Variety for Threescore Years, besides her Childhood, was Twelve Year a Whore, five times a Wife (whereof once to her brother), Twelve Year a Thief, Eight Year a Transported Felon in Virginia, at last grew Rich, liv'd Honest, and died a Penitent*

Bonus:

Ways of Seeing is a 1972 work of art criticism by John Berger based on his TV series that was a radical counter-argument to Kenneth Clark's *Civilisation*.

QUIZ 2

ROUND I

QUICKFIRE:
THE CHRISTMASAURUS

Six questions about books and Christmas . . .

I.

Which much-loved book of 1959 contains a description of Christmas in the Gloucestershire village of Slad soon after the First World War?

2.

In Kenneth Grahame's *The Wind in the Willows*, which of the characters/animals returns home for Christmas where he and another of the characters are serenaded by field mice?

3.

'I'm not going to tell you my whole goddamn autobiography or anything. I'll just tell you about this madman stuff that happened to me around last Christmas,' says the narrator Holden Caulfield – at the beginning of which 1951 novel?

37

4.

Two of the four New Testament gospels don't contain any mention at all of the birth of Jesus – please name either of them.

5.

Which poet, born on Christmas Eve 1822, wrote the poems 'Dover Beach', 'The Scholar Gipsy' and 'Thyrsis', which contains the famous reference to Oxford's 'dreaming spires'?

6.

'And what rough beast, its hour come round at last,/ Slouches towards Bethlehem to be born?' are the last lines of the poem 'The Second Coming' by which poet – who in 1923 became the first Irish writer to win the Nobel Prize in Literature?

*And six where the answers
contain a word associated with Christmas . . .*

7.

In 2006 Orhan Pamuk became the first writer from which country to win the Nobel Prize in Literature?

38

8.

What was E. M. Forster's first novel?

9.

The adult novels of which writer – all of which were published after she was 70 – include *The Vacillations of Poppy Carew*, *Harnessing the Peacocks* and *The Camomile Lawn*?

10.

Who wrote the bestselling 2002 novel about the Irish famine *Star of the Sea*?

11.

A Bone in My Flute is the autobiography of which musician, the former lead singer of Frankie Goes to Hollywood?

12.

Detox for Life and *How to Do Extreme Sudoku* are two of the books by which former star of the TV show *Countdown*?

ANSWERS ON PP. 57–8

ROUND 2

NAME THE AUTHOR

*Can you guess the writer from these clues
(and, of course, the fewer you need the better)?*

I.

A. Her grandfather wrote what appears to be, from all available records, the bestselling British novel of the nineteenth century.

B. She was of French Huguenot descent.

C. She hated Alfred Hitchcock's film adaptation of her story 'The Birds'.

D. Hitchcock also made two other films based on her books: *Jamaica Inn* and *Rebecca*.

40

2.

A.   In 1996 Daniel Day-Lewis became his son-in-law.

B.   He wrote the 1961 film *The Misfits*, starring his then wife.

C.   One of his plays depicts the Loman family.

D.   Another shares its name with a famous snooker venue.

3.

A.   His first wife wrote the possibly autobiographical 1962 novel *The Pumpkin Eater* about a marriage being undermined by the husband's philandering.

B.   As a lawyer, he defended a record shop manager who'd been charged with obscenity for displaying the album *Never Mind the Bollocks, Here's the Sex Pistols*.

C.   His most famous character had a wife whose name was Hilda but who was usually referred to as 'She Who Must Be Obeyed'.

D.   The character in question was Horace Rumpole.

ANSWERS ON PP. 59–61

4.

A. She once turned down a proposal of marriage from a man named Harris Bigg-Wither.

B. Mark Twain said that he often wanted to criticise her, but 'her books madden me so that I can't conceal my frenzy from the reader; and therefore I have to stop every time I begin'.

C. She had a sister called Cassandra.

D. Her novel *Persuasion* was published posthumously.

5.

A. Between 1956 and 1958 he was a teacher at Eton.

B. His father was a conman whose life he drew on for a 1986 book described by Philip Roth as 'the best English novel since the war'.

C. The main character in his breakthrough novel is Alec Leamas.

D. His real name is David Cornwell.

ANSWERS ON PP. 59–61

6.

A. He shares a name with 'the greatest character in TV history', according to both Channel 4 viewers in Britain and *Entertainment Weekly* magazine in America.

B. One of his works was the first ever Penguin Classic – and for many years Penguin's bestselling book.

C. He might not have existed.

D. That first Penguin Classic was a kind of sequel to *The Iliad*.

43

ROUND 3

EXTRACTS: IT'S ALL
ABOUT THE CHILDREN

Can you identify these novels from their young narrators?
One point for the title, and another for the author in each case.

I.

From a novel of 1960:

Maycomb was an old town, but it was a tired old town when I first knew it. In rainy weather the streets turned to red slop; grass grew on the sidewalks, the courthouse sagged in the square. Somehow, it was hotter then: a black dog suffered on a summer's day; bony mules hitched to Hoover carts flicked flies in the sweltering shade of the live oaks on the square. Men's stiff collars wilted by nine in the morning. Ladies bathed before noon, after their three-o'clock naps, and by nightfall were like soft tea-cakes with frostings of sweat and sweet talcum . . .

When I was almost six and Jem was nearly ten, our summertime boundaries . . . were Mrs Henry Lafayette Dubose's house two doors to the north of us, and the

Radley Place three doors to the south. We were never tempted to break them. The Radley Place was inhabited by an unknown entity the mere description of whom was enough to make us behave for days on end; Mrs Dubose was plain hell.

Q2

2.

From a novel of 2003:

This will not be a funny book. I cannot tell jokes because I do not understand them. Here is a joke, as an example. It is one of Father's.

His face was drawn but the curtains were real.

I know why this is meant to be funny. I asked. It is because *drawn* has three meanings, and they are 1) drawn with a pencil, 2) exhausted, and 3) pulled across a window, and meaning 1 refers to both the face and the curtains, meaning 2 refers only to the face, and meaning 3 refers only to the curtains.

If I try to say the joke to myself, making the word mean the three different things at the same time, it is like hearing three different pieces of music at the same time which is uncomfortable and confusing and not nice like white noise. It is like three people trying to talk to you at the same time about different things.

And that is why there are no jokes in this book.

45

3.

From a novel of 1884:

Q2

Her sister, Miss Watson, a tolerable slim old maid, with goggles on, had just come to live with her, and took a set at me now, with a spelling-book. She worked me middling hard for about an hour, and then the widow made her ease up. I couldn't stood it much longer. Then for an hour it was deadly dull, and I was fidgety. Miss Watson would say, 'Don't put your feet up there . . . why don't you try to behave?' Then she told me all about the bad place, and I said I wished I was there. She got mad then, but I didn't mean no harm. All I wanted was to go somewheres; all I wanted was a change, I warn't particular. She said it was wicked to say what I said; said she wouldn't say it for the whole world; *she* was going to live so as to go to the good place. Well, I couldn't see no advantage in going where she was going, so I made up my mind I wouldn't try for it. But I never said so, because it would only make trouble, and wouldn't do no good.

ANSWERS ON PP. 62–4

4.

*From a novel of 1954 – although for this one,
you only need the novel series and the author:*

Another thing about xmas eve is that your pater always reads the xmas carol by c. dickens. You canot stop this aktualy although he pretend to ask you whether you would like it. He sa:

Q2

Would you like me to read the xmas carol as it is xmas eve, boys?

We are listening to the space serial on the wireless, daddy . . .

Then he rub hands together and sa You will enjoy this boys it is all about ghosts and goodwill. It is tip-top stuff . . . Nothing in the world in space is ever going to stop those fatal words:

Marley was dead.

Personaly I do not care a d. whether Marley was dead or not it is just that there is something about the xmas Carol which makes paters and grown-ups read with grate XPRESION, and this is very embarassing for all . . . When Tiny Tim sa God bless us every one your pater is so overcome he burst out blubbing.

ROUND 4

ORDER, ORDER

*Can you place these three items in ascending, correct or
their usual order?*

1.

A. _____ *of Champions* (Kurt Vonnegut novel)

B. _____ *at the Homesick Restaurant* (Anne Tyler novel)

C. *Naked* _____ (William S. Burroughs novel)

2.

A. A 1980 novel by J. L. Carr

B. Aleksandr Solzhenitsyn's first novel, about a political
    prisoner

C. Peter Mayle's France-based non-fiction bestseller of
    the 1980s

3.

A.   Roald Dahl's crocodile, according to the title of a 1978 book

B.   David Lodge's world, according to the title of a 1984 novel

C.   Raymond Chandler's first novel

Q 2

4.

A.   The creator of Jack Reacher

B.   The fourth Bridget Jones book

C.   Anne Catherick in the title of a classic Victorian thriller by Wilkie Collins

5.

A.   *The _____ and I* (a 1992 novel by Sue Townsend)

B.   The novel in which H. Rider Haggard's character Allan Quatermain first appears

C.   The Tom Clancy character who starts in the CIA and ends up as US president

ANSWERS ON PP. 65–7

6.

A. The Nevil Shute novel that has an abbreviated Australian place name in the title

Q2

B. The Paul Du Noyer book about London music whose title is the same as that of The Jam's first single

C. Gertrude's son in a play by Shakespeare

ROUND 5

QUICKFIRE: BLOOD FEVER

Q2

Family time . . .

1.

Who wrote *The Tenant of Wildfell Hall*, featuring a violent drunkard partly based on her brother Branwell?

2.

Which bestselling novel by Emma Healey – the winner of the 2014 Costa First Novel Award – was inspired by a remark of her grandmother's, who was suffering from dementia?

3.

Which TV presenter – and former drag artist – wrote the bestselling 2008 memoir *At My Mother's Knee . . . and Other Low Joints*?

51

4.

Which novel begins (in translation): 'Mother died today. Or maybe yesterday, I don't know'?

5.

What's the surname of the fictional family, created by John Galsworthy, whose members include Jolyon, Fleur, Irene and Soames?

6.

Who won the 2007 Man Booker Prize with *The Gathering*, her novel about a large Irish family?

7.

Which American writer traced his African ancestors in the book *Roots*?

8.

Which brothers – better known for their work in another field of literature – were the first compilers of the German equivalent of the *Oxford English Dictionary*?

ANSWERS ON PP. 68–9

9.

In a play by Sophocles, who killed his father and married his mother?

10.

In 1990 the British Crime Writers' Association voted whose 1951 book *The Daughter of Time* the greatest ever crime novel?

11.

Who imagined a world in which Germany had won the Second World War in his novel *Fatherland*?

12.

In which children's classic by Clive King does a boy discover and befriend a caveman while staying with his grandparents?

ANSWERS ON PP. 68–9

ANSWERS
TO QUIZ 2

ROUND I

1.

Cider with Rosie – by Laurie Lee

2.

Mole

3.

The Catcher in the Rye – by J. D. Salinger

4.

Mark or John

5.

Matthew Arnold

6.

W. B. Yeats

7.

Turkey

8.

Where Angels Fear to Tread

9.

Mary Wesley (who had written three children's books by then, starting in her late fifties)

10.

Joseph O'Connor

11.

Holly Johnson

12.

Carol Vorderman

Bonus:

The Christmasaurus is a bestselling children's book of 2016 by Tom Fletcher about a boy and the dinosaur he meets one Christmas Eve. It's since been adapted for both stage and screen.

ROUND 2

1.

Daphne du Maurier, whose Huguenot background explains that French surname – and who apparently disliked Hitchcock's decision to relocate 'The Birds' from Cornwall to California. (Incidentally, another of her short stories, 'Don't Look Now', was the basis of another classic film.) As for that nineteenth-century bestseller, it was *Trilby* by George du Maurier, who also drew the Punch cartoon that gave us the phrase 'a curate's egg'. His tale of Trilby O'Ferrall – an artists' model living in Paris during its bohemian pomp – is less read now, but its influence lives on in several ways. Trilby's sinister mentor is called Svengali; her euphemism for posing nude is 'in the all together'; the trilby hat is named after the one she wore in Beerbohm Tree's wildly successful stage version and soon became a craze in itself. The book was such a hit in America too that there's a town in Florida named after it.

2.

Arthur Miller, whose daughter Rebecca married Day-Lewis; the couple met when he came to Miller's house while working on a film version of *The Crucible* (the play that shares its name with a snooker venue). Willy Loman and his long-suffering family are in *Death of a Salesman* – and *The Misfits* starred Marilyn Monroe, whom Miller had married in 1956, leading to the celebrated *Variety* magazine headline 'Egghead Weds Hourglass'. It proved to be her last completed film – as well as Clark Gable's. He died less than two weeks after filming was finished.

3.

John Mortimer, whose first wife, Penelope, wrote *The Pumpkin Eater*, later a Penguin Modern Classic. Mortimer won the Sex Pistols case with the aid of an expert academic witness who argued that 'bollocks' was a venerable English word for 'nonsense'. By 2019, you may remember, it could be used by the Lib Dems as a slogan in their European election campaign. Mortimer also successfully defended Hubert Selby Jr's *Last Exit to Brooklyn* against obscenity charges.

4.

Jane Austen, who was 27 (i.e. positively an old maid) when she rejected Bigg-Wither's proposal, having accepted the night before. Twain's dislike of her also led him to say, 'Every time I read *Pride and Prejudice* I want to dig her up and beat her over the skull with her own shin-bone!' Cassandra was her confidante as well as her sister – and the recipient of letters that have sometimes disconcerted Austen fans with the unkindness of the jokes they contain. Of a woman who'd had a miscarriage, she wrote: 'Mrs Hall was brought to bed yesterday of a dead child, some weeks before she expected, owing to a fright. I suppose she happened unawares to look at her husband.'

5.

John le Carré. Alec Leamas is the protagonist of *The Spy Who Came in from the Cold*, the success of which allowed le Carré to leave MI6 and write full-time. He needed a pseudonym because foreign office officials were forbidden from publishing under their own name. The novel so admired by Roth was *A Perfect Spy*.

6.

Homer – the TV character being Homer Simpson. The first Penguin Classic, *The Odyssey*, was translated by E. V. Rieu in 1946, and remained Penguin's bestselling title until the publication of the uncensored *Lady Chatterley's Lover* fourteen years later.

ROUND 3

I.

From *To Kill a Mockingbird* by Harper Lee – where the young narrator is Scout Finch, growing up with her brother, Jem, in Maycomb, Alabama, a place closely modelled on Lee's own hometown of Monroeville. The 'unknown entity' at the Radley Place is the reclusive Boo Radley, who later plays a key role – and later still gave the Britpop band The Boo Radleys their name. *To Kill a Mockingbird* was Lee's first novel, and after winning the Pulitzer Prize it went on to become one of the bestselling of all time. Even so, Lee didn't publish so much as a word of fiction again. Until, that is, in 2015, when she was 89 and a second novel, *Go Set a Watchman*, was published to great excitement and enormous sales. Opinion remains divided, however, as to whether it really was a separate work or just a first draft of *To Kill a Mockingbird*. To the great distress of some readers, it also portrayed Scout's father Atticus as far less unfailingly noble (and far more racist) than in *Mockingbird*.

2.

From Mark Haddon's *The Curious Incident of the Dog in the Night-Time*. The narrator is Christopher Boone – or as he prefers to call himself Christopher John Francis Boone – a fifteen-year-old boy with Asperger's syndrome, a condition which has effects, such as not getting jokes, that he can describe but can't do anything about. The book was also adapted into a West End play, which made the news in December 2013 when the ceiling of the theatre collapsed, injuring around 80 people. Its title comes from the Sherlock Holmes story 'The Adventure of the Silver Blaze', in which Holmes draws the attention of a Scotland Yard detective to 'the curious incident of the dog in the night-time'. 'The dog did nothing in the night-time,' says the detective. 'That was the curious incident,' replies Holmes.

3.

From *The Adventures of Huckleberry Finn* by Mark Twain – with Huck, as ever, resisting all well-meaning attempts to 'sivilize' him, in this case by his guardian, the Widow Douglas, and her sister, Miss Watson. Twain's determination to use authentic American speech rather than impersonate European novels meant that in 1935 Ernest Hemingway famously declared: 'All modern American literature comes from one book by Mark Twain called *Huckleberry Finn*. It's the best book we've had.'

A 2

4.

From the *Molesworth* series by Geoffrey Willans, with illustrations by Ronald Searle. (And special congratulations if you happened to know it was from the second book in the series, *How to Be Topp*.) Nigel Molesworth, celebrated pupil of St Custard's and coiner of the phrase 'as any fule kno', was firmly of the belief that Christmas is a time for grownups – and a bit of a strain for children, what with having to pretend everything is a surprise. He may also have slightly missed the point of *A Christmas Carol*, considering Tiny Tim 'a weed' and the unreformed Scrooge the best character in fiction 'next to tarzan of the apes'.

ROUND 4

1.

A, C, B – meal times:

<u>Breakfast</u> of Champions

Naked <u>Lunch</u>

<u>Dinner</u> at the Homesick Restaurant

2.

B, A, C – measures of time in ascending order:

<u>One Day</u> in the Life of Ivan Denisovich

<u>A Month</u> in the Country
(about a damaged First World War survivor)

<u>A Year</u> in Provence

3.

B, C, A – ascending order of size:

Small World

The Big Sleep (the book that introduced the world to Chandler's detective Philip Marlowe)

The Enormous Crocodile

4.

B, A, C – ascending order of age:

Bridget Jones's Baby, by Helen Fielding

Lee *Child* (Reacher, a former American military policeman, first appeared in 1997's *Killing Floor* and Child has written at least one novel a year about his adventures since then)

The Woman in White

5.

C, A, B – jack, queen, king:

Jack Ryan

The Queen and I (in which, following the election of a Republican government, the royal family have to live on a council estate)

King Solomon's Mines

6.

*C, A, B – human settlements in
ascending order of size:*

<u>Hamlet</u>

A <u>Town</u> Like Alice (as in Alice Springs)

In the <u>City</u>

ROUND 5

1.

Anne Brontë

2.

Elizabeth Is Missing. (Her grandmother's remark was 'My friend is missing'.)

3.

Paul O'Grady – once better known as Lily Savage

4.

The Outsider (or *L'Étranger*) – by Albert Camus

5.

Forsyte – in *The Forsyte Saga*

6.

Anne Enright

7.

Alex Haley

8.

The Brothers Grimm, Jacob and Wilhelm

9.

Oedipus (in his defence he didn't realise they *were* his father and mother)

10.

Josephine Tey

11.

Robert Harris

12.

Stig of the Dump

Bonus:

Blood Fever is a novel in the *Young Bond* series by Charlie Higson, which features James Bond in his already adventurous days as an Eton schoolboy.

QUIZ 3

ROUND I

QUICKFIRE:
ANOTHER COUNTRY

Around the world in twelve questions . . .

I.

What country is referred to in the title of Alan Paton's *Cry, the Beloved Country*?

2.

Which 1990s bestseller by Peter Høeg features a heroine who grew up in Greenland?

3.

Who created the geographically named characters Orinoco, Wellington, Madame Cholet and Great Uncle Bulgaria?

4.

What's the city in Armistead Maupin's *Tales of the City* series?

5.

What's the name of the memoir by Karen Blixen that was adapted into a film, starring Meryl Streep as Blixen, which won the Best Picture Oscar in 1986?

6.

Where does a tree grow according to the title of a novel by Betty Smith, a much-loved book in America since it was published in 1943?

7.

As what is Sarah Woodruff known in the title of a book by John Fowles?

8.

The title of which 1973 novel by J. P. Donleavy was later used as the name of a celebrated Christmas song?

9.

Aldous Huxley's *Eyeless in Gaza* takes its title from a phrase in the seventeenth-century poem *Samson Agonistes* by which poet – who was himself blind when he wrote it?

10.

Great Fortune, Spoilt City and *Friends and Heroes* are the three novels in which trilogy by Olivia Manning?

11.

In which children's book by Erich Kästner do the characters of the title follow the baddie Max Grundeis around Berlin?

Q3

12.

Complete the title of the 1973 Booker-winning novel by J. G. Farrell, set during the Indian Mutiny/Rebellion of 1857: *The Siege of . . .*

75

ROUND 2

NAME THE AUTHOR

*Can you guess the writer from these clues
(and, of course, the fewer you need the better)?*

I.

A.   She published her first book while working as a
gossip columnist for the *Daily Express*.

B.   A typical piece of advice that she offered to married
women was: 'A woman should say: "Have I made
him happy? Is he satisfied? Does he love me more
than he loved me before? Is he likely to go to bed
with another woman?" If he does, then it's the
wife's fault because she is not trying to make him
happy.'

C.   She had the longest ever entry in *Who's Who* –
largely because it listed all of her books.

D.   Her daughter was Princess Diana's stepmother.

76

2.

A. His brother Peter, a travel writer, was married to the actress Celia Johnson, star of the film *Brief Encounter*.

B. The protagonist of fourteen of his books was named after the author of a guide to Caribbean birds.

C. His only children's book was *Chitty-Chitty-Bang-Bang*.

D. When he died in 1964, he left the novel *The Man with the Golden Gun* unfinished.

3.

A. He had a pet raven called Grip.

B. The main character in his favourite of his own novels had the same initials as him, only in reverse.

C. He was the first novelist to appear on a British banknote.

D. The only one of his novels without any scenes set in London is *Hard Times*.

Q3

4.

A.   Her first book sold only two copies, which – given that it was a joint collection of verse by her and two other poets – was one more than the number of people who wrote it.

B.   Her only novel begins: 'I have just returned from a visit to my landlord' . . .

C.   . . . and is one of the biggest ever selling Penguin Classics.

D.   The name of the landlord is Heathcliff.

5.

A.   He wrote the screenplay for the James Bond film *You Only Live Twice*.

B.   In 2010 his granddaughter married the singer Jamie Cullum.

C.   He was named after a Norwegian explorer.

D.   His first book for children was *The Gremlins* in 1943 and his second, nearly 20 years later, was *James and the Giant Peach*.

6.

A. He was known for three years as Sebastian Melmoth.

B. One of his early plays was banned in Britain because it featured biblical characters.

C. His most famous play opens 'in Algernon's flat in Half-Moon Street'.

D. In 1895 he was sentenced to two years' hard labour in prison.

ROUND 3

EXTRACTS:
BOOK OF THE YEAR

All these passages come from the bestselling novel in America of a particular year (according to the magazine Publishers Weekly*). In each case, there's one point for the novel and another for the author.*

I.

From America's bestselling novel of 1969:

It alleviates nothing fixing the blame – blaming is still ailing, of course, of course – but nonetheless, what *was* it with these Jewish parents, *what*, that they were able to make us little Jewish boys believe ourselves to be princes on the one hand, unique as unicorns on the one hand, geniuses and brilliant like nobody has ever been brilliant and beautiful before in the history of childhood – saviours and sheer perfection on the one hand, and such bumbling, incompetent, thoughtless, helpless, selfish, evil little . . . *ingrates*, on the other!

ANSWERS ON PP. 98–9

2.

From America's bestselling novel of 1939:

To the red country and part of the gray country of Oklahoma, the last rains came gently, and they did not cut the scarred earth. The plows crossed and recrossed the rivulet marks. The last rains lifted the corn quickly and scattered weed colonies and grass along the sides of the roads so that the gray country and the dark red country began to disappear under a green cover. In the last part of May the sky grew pale and the clouds that had hung in high puffs for so long in the spring were dissipated. The sun flared down on the growing corn day after day until a line of brown spread along the edge of each green bayonet.

3.

From America's bestselling novel of 2016
(although it's by a British writer):

The pre-mixed gin and tonic fizzes up over the lip of the can as I bring it to my mouth and sip. Tangy and cold, the taste of my first ever holiday with Tom, a fishing village on the Basque coast in 2005. In the mornings we'd swim the half-mile to the little island in the bay, make love on secret hidden beaches; in the afternoons we'd sit at a bar drinking strong, bitter gin and tonics, watching swarms of beach footballers playing chaotic 25-a-side games on the low-tide sands.

81

I take another sip, and another; the can's already half empty but it's OK, I have three more in the plastic bag at my feet. It's Friday, so I don't have to feel guilty about drinking on the train. TGIF. The fun starts here.

4.

From America's bestselling novel of 1958 (although it's by a Russian writer):

Yura looked around and saw the same things that had caught Lara's eye not long before. Their sleigh raised an unnaturally loud noise, which awakened an unnaturally long echo under the ice-bound trees of the gardens and boulevards. The frosted-over windows of houses, lit from inside, resembled precious caskets of laminated smoky topaz. Behind them glowed Moscow's Christmas life, candles burned on trees, guests crowded, and clowning mummers played at hide-and-seek and pass-the-ring.

ANSWERS ON PP. 98–9

ROUND 4

ODD ONE OUT

*In each of these batches of four,
what item doesn't belong and why?*

I.

A. The final book in C. S. Lewis's *The Chronicles of Narnia*

B. The 1898 novel in which the Martians land near Woking

C. A treatise by Sun Tzu from the fifth century BC that's still in use today

D. Joey in a novel by Michael Morpurgo that became both a play and a Steven Spielberg film

2.

A. David Attenborough's book about reptiles and amphibians that accompanied a television series of the same name

B. *Love in a* _____ *Climate* (Nancy Mitford's 1949 follow-up to *The Pursuit of Love*)

C. *The* _____ *of the Day* (Elizabeth Bowen's 1948 novel set around the London blitz)

D. Stella Gibbons's 1932 novel about the Starkadder family

3.

A. Edith Wharton's deceptively cheerful-sounding breakthrough novel

B. Anita Brookner's 1984 Booker Prize-winner

C. Isabel Allende's bestselling first novel

D. Henrik Ibsen's once scandalous play in which Nora Helmer leaves her husband and children

ANSWERS ON PP. 100–102

4.

A. V. S. Naipaul's breakthrough novel

B. The murderer of Sir Danvers Carew in a book by Robert Louis Stevenson

C. The books by Roger Hargreaves that are mashed up with *Doctor Who* in a series of Puffin books launched in 2017

D. Beatrix Potter's eponymous hedgehog

5.

A. Michael Lewis's book about the economic crisis of 2007–8 that, unusually for a book on finance, went on to become a hit movie starring Christian Bale, Ryan Gosling, Steve Carell and Brad Pitt

B. Donna Tartt's second novel

C. The 1943 novella by Antoine de Saint-Exupéry that's now one of the biggest selling books in the history of the world

D. The E. B. White character born to human parents in New York who's perfectly normal – except for being four inches high and looking like a mouse

ANSWERS ON PP. 100–102

6.

A. The short story by James Thurber about a fantasist, whose name is now a byword for fantasists everywhere

B. The Kate Atkinson novel that won the 2013 Costa Novel Award

C. The book by James Boswell that's never been out of print since it was published in 1791

D. The Thomas Mann novel in which a distinguished composer becomes fatefully obsessed with a beautiful boy

ANSWERS ON PP. 100–102

ROUND 5

QUICKFIRE: AT HOME

Q3

Around the UK in twelve questions . . .

I.

What British place name follows *Tropic of* in the title of a 1974 novel by Leslie Thomas?

2.

Which village in Devon – named after an 1855 novel by Charles Kingsley – is the only British place name that has an exclamation mark?

3.

A Romance of Exmoor is the subtitle of which 1869 novel by R. D. Blackmore?

4.

Octavia Pole, Deborah Jenkyns and Mary Smith are among the inhabitants of which fictional British town in a book by Elizabeth Gaskell?

5.

In a novel by John Wyndham, what's the name of the fictional British village where all the fertile women suddenly become pregnant with what turn out to be strange golden-eyed alien children?

6.

Which Thomas Hardy heroine is arrested for murder at Stonehenge?

7.

Which postcode area gives its title to a 2012 novel by Zadie Smith?

8.

As what is Charles Primrose known in the title of a novel by Oliver Goldsmith?

9.

Which city is the setting for James Kelman's Booker Prize-winner *How Late It Was, How Late*?

10.

Which book by Richard Llewellyn, set in South Wales, was America's bestselling novel of 1940 – and the following year was adapted into a film that won the Oscar for Best Picture?

Q3

11.

Which long poetic work – and cornerstone of English literature – begins at the Tabard Inn in Southwark?

12.

Who wrote the Yorkshire-based novel *South Riding*, adapted for the screen several times – including by Andrew Davies in 2011, with Anna Maxwell Martin and David Morrissey among the cast?

ANSWERS
TO QUIZ 3

ROUND I

1.

South Africa

2.

Miss Smilla's Feeling for Snow

3.

Elisabeth Beresford (those being the names of various Wombles)

4.

San Francisco

5.

Out of Africa

6.

Brooklyn – the book's title therefore being *A Tree Grows in Brooklyn*

7.

The French Lieutenant's Woman

8.

A Fairy Tale of New York

9.

John Milton

10.

The Balkan Trilogy

11.

Emil and the Detectives

12.

Krishnapur (Krishnapur itself is a fictional city)

Bonus:

Another Country is a 1962 novel by James Baldwin about the repercussions that follow the death of an African-American jazzman in Greenwich Village, New York.

ROUND 2

1.

Barbara Cartland – mother of Raine Spencer – who apparently published 723 books before her death in 2000; in her heyday at the rate of one a fortnight. 'The only books Diana ever read were mine,' she said after the Princess's death, 'and they weren't awfully good for her.'

2.

Ian Fleming, who had a house called Goldeneye in Jamaica – where *Birds of the West Indies* (1936) by James Bond was a favourite guidebook of his. He chose the name for his hero, he said, because he wanted a plain-sounding one and James Bond was 'the dullest I ever heard'. In the film *Die Another Day*, Pierce Brosnan as Bond is seen reading the book – although the author's name is discreetly hidden.

3.

Charles Dickens, who was on the £10 note from 1992 to 2003. *Hard Times* is set in the fictional Coketown in the industrial north – and the favourite of his own novels was

David Copperfield. (As he wrote in a preface to the book near the end of his life: 'I am a fond parent to every child of my fancy . . . But, like many fond parents, I have in my heart of hearts a favourite child. And his name is DAVID COPPERFIELD.')

4.

A3

Emily Brontë. That first book was *Poems* by Currer, Ellis and Acton Bell, the pseudonyms for Charlotte, Emily and Anne Brontë in descending order of their ages. That rather more successful novel was *Wuthering Heights*, which was her only novel mainly because she died at the age of 30.

5.

Roald Dahl, whose Norwegian parents named him after Roald Amundsen. (In 1980, Dahl had an asteroid named in his honour by the Czech astronomer Antonín Mrko.) Between *The Gremlins* and *James and the Giant Peach* Dahl wrote for adults, mostly the short stories that were later collected as *Tales of the Unexpected*. The granddaughter who married Jamie Cullum is Sophie, the inspiration for the heroine of *The BFG* – and later a model. Dahl, incidentally, also wrote the screenplay for Ian Fleming's aforementioned only book for children, *Chitty-Chitty-Bang-Bang*, although the film's title was unhyphenated – by no means the least of the changes to the novel, where for one thing there's no mythical child-hating land called Vulgaria.

6.

Oscar Wilde. That most famous play is *The Importance of Being Earnest*, which begins in the flat of Algernon Moncrieff, friend of the main character, Jack 'Ernest' Worthing. The banned one was *Salomé*, written in French for Sarah Bernhardt to perform in London in 1892. Because of the ban, though, it didn't have its premiere until 1896 in Paris, by which time Wilde was serving his prison sentence for 'sodomy and gross indecency'. After his release he lived in exile in Paris, under the name Sebastian Melmoth.

ROUND 3

I.

Portnoy's Complaint by Philip Roth, which takes the form of an extended monologue by Alexander Portnoy to his psychoanalyst Dr Spielvogel. As well as doing much to establish – or at least popularise – the idea of Jewish mothers' oppressiveness, the book was also famous for its sexual frankness. (My ellipsis in that passage discreetly covers one of the milder swear words.) Among other consequences, this led to it being declared a 'prohibited import' in Australia.

2.

The Grapes of Wrath by John Steinbeck – with Oklahoma being afflicted by the dustbowl that will cause the Joad family to head for California, with not entirely happy results.

3.

The Girl on the Train by Paula Hawkins – that part narrated by the alcoholic Rachel Watson, who's also the girl of the title. Hawkins's novel was also America's third bestselling novel of 2015 – and it didn't do badly in Britain either, where it was also at number three for 2015 and lost out narrowly in 2016 only to *Harry Potter and the Cursed Child*.

A3

4.

Doctor Zhivago by Boris Pasternak – Yura is Zhivago's first name and Lara the woman he falls for (hence the famous 'Lara's Theme' in the film version). *Doctor Zhivago* was also America's second biggest selling novel of 1959, behind Leon Uris's *Exodus*.

ROUND 4

I.

The odd one out is **A**, *The Last Battle* – because the others feature full-scale war. It's in H. G. Wells's *The War of the Worlds* that the Martians begin their invasion of the Earth near Woking, where Wells was living at the time. The tactics in Sun Tzu's *The Art of War* helped Chairman Mao to win the civil war in China – while General Giap used them to beat the French and then the Americans in Vietnam. (The book is still recommended reading for US Marines and apparently inspired Brazil's 2002 World Cup-winning team.) Joey is the eponymous equine main character in Morpurgo's *War Horse*.

2.

The odd one out is **C**, *The Heat of the Day* – the others all being cold: Attenborough's *Life in Cold Blood*; Mitford's *Love in a Cold Climate*; and Gibbons's parody of rural novels *Cold Comfort Farm*, the book that gave us the phrase 'something nasty in the woodshed'.

3.

The odd one out is **B**, because that's _Hotel_ du Lac – while the others are mere houses. Wharton's _The House of Mirth_ is definitely not as jolly as it sounds, although that won't come as a surprise to people who know their Bible: 'The heart of the wise is in the house of mourning; but the heart of fools is in the house of mirth,' says the ever-cheerful Book of Ecclesiastes. Isabel Allende launched her fiction-writing career with _The House of the Spirits_, and Nora is in Ibsen's _A Doll's House_. According to one critic, Nora's slamming of the door when she walks out for good 'reverberated across the roof of the world' – and a woman leaving her family was considered so shocking that the first productions in both Germany and Britain went for a 'happier' ending in which Nora decides to stay.

A3

4.

This is a Mr and Mrs question with **D** as the odd one out, because that's _Mrs Tiggy-Winkle_. The others are Naipaul's _A House for Mr Biswas_; _Mr_ Hyde in Stevenson's _The Strange Case of Dr Jekyll and Mr Hyde_; and Roger Hargreaves's _Mr Men_. The Puffin series, by Hargreaves's son Adam, has a different book for each incarnation of the Doctor – which meant that when Jodie Whittaker took over the role he needed to mash up _Doctor Who_ with Hargreaves's _Little Miss_ series instead.

5.

The odd one out is **A**, *The Big Short*. The remaining three were little in various ways: Tartt's *The Little Friend*; Saint-Exupéry's *The Little Prince*; and White's Stuart *Little* in the book of the same name.

6.

A question of life and death, with **D** as the odd one out, because that's *Death in Venice*. The others are Thurber's *The Secret Life of Walter Mitty*; Atkinson's *Life after Life*; and Boswell's *The Life of Samuel Johnson*.

ROUND 5

I.

Ruislip

2.

Westward Ho! (Kingsley's novel, set in nearby Bide-ford, was a bestseller – and a local peer saw a tourist opportunity.)

3.

Lorna Doone

4.

Cranford (also the title of the novel)

5.

Midwich (in *The Midwich Cuckoos*)

6.

Tess (Durbeyfield) in *Tess of the D'Urbervilles*

7.

NW

8.

The Vicar of Wakefield

9.

Glasgow

10.

How Green Was My Valley

11.

The Canterbury Tales – by Geoffrey Chaucer

12.

Winifred Holtby

Bonus:

At Home: A Short History of Private Life is a non-fiction book by Bill Bryson about how our houses ended up the way they have (and plenty more besides).

A3

QUIZ 4

ROUND I

QUICKFIRE:
THE WILL TO POWER

*Six questions where the answers all contain the surname of
a British prime minister . . .*

1.

The Diana Chronicles is a book about the former Princess
of Wales by which former editor of *The New Yorker* and
Vanity Fair?

2.

In Mark Twain's *The Adventures of Tom Sawyer*, what's the
name of the classmate that Tom falls in love with?

3.

What's John Steinbeck's longest novel, which takes its
title from a phrase in the Bible and became a film starring
James Dean?

4.

Barry Humphries's autobiography *More Please* chronicles his childhood in which city?

5.

What name follows *The Miseducation of* . . . in the title of a lesbian coming-of-age novel by Emily M. Danforth that became the basis of an indie film in 2018?

6.

Which African-American author's works include *Giovanni's Room*, *Notes of a Native Son* and *Go Tell It on the Mountain*?

And six where the answers all contain the surname of an American president . . .

7.

Which British author's works include *Wise Children*, *Nights at the Circus* and *The Bloody Chamber and Other Stories*?

8.

Who wrote *A Confederacy of Dunces*, published to great acclaim eleven years after he'd committed suicide – partly because he couldn't find a publisher for the novel?

ANSWERS ON PP. 127–8

9.

Who, in 1978, became the first woman to have a wholly self-penned number one single in Britain, with a song based on a novel of 1847?

10.

Whose 1999 childcare guide *The Contented Little Baby Book* somewhat controversially advocates a strict feeding and sleeping routine for babies?

Q4

11.

The History of Rasselas, Prince of Abissinia was the only novel by which eighteenth-century critic, poet and lexicographer?

12.

Who wrote the true-crime book *In Cold Blood*, often called the first non-fiction novel?

III

ANSWERS ON PP. 127–8

ROUND 2

NAME THE AUTHOR

*Can you guess the writer from these clues
(and, of course, the fewer you need the better)?*

Q4

1.

A. He was Britain's biggest selling non-fiction writer of the 2000s.

B. He shares a catchphrase with the main character of a much-loved British sitcom.

C. His surname is a male first name.

D. His *5 Ingredients* was Britain's biggest selling book of 2017.

ANSWERS ON PP. 129–31

2.

A. Now one of Britain's most famous poets, he died in obscurity in London in 1827.

B. He had visions of angels from boyhood onwards.

C. He claimed that in 1788 his brother Robert taught him the technique of 'illuminated printing' that he used to produce many of his books – even though Robert had died the previous year.

D. One of his poems is sung at every Labour Party conference and on every day of English Test cricket – and is the unofficial anthem of the Women's Institute.

Q4

3.

A. A passionate opponent of slavery, she worked as a nurse in the American Civil War.

B. Her middle name is a month of the year.

C. She wrote a series of novels about the March family . . .

D. . . . which begins with *Little Women*.

ANSWERS ON PP. 129–31

4.

A. He served as an artillery officer in the Crimean War.

B. In later life he became such a radical Christian that he was excommunicated by the Orthodox Church.

C. Parts of one of his novels first appeared in the magazine *The Russian Messenger* under the title *The Year 1805*.

D. Its final title was *War and Peace*.

5.

A. His middle name was Hoyer.

B. His only play was *Buchanan Dying*, about James Buchanan, the only US president from Pennsylvania.

C. His first big seller, in 1968, was *Couples*, which led to him appearing on the cover of *Time* magazine with the strapline 'The Adulterous Society'.

D. He wrote four novels about a former high-school basketball star called Harry Angstrom, better known as 'Rabbit'.

ANSWERS ON PP. 129–31

6.

A. Before becoming a full-time writer in her forties, she was a TV executive.

B. Her real name is Erika Leonard.

C. The three bestselling novels in Britain of 2012 were all by her . . .

D. . . . and all featured two main characters whose first names are Anastasia and Christian.

Q4

115

ROUND 3

EXTRACTS:
PLEASED TO MEET YOU

Q4

*Can you identify the four celebrated fictional characters
making their first appearances here for a point – plus
the book and author in each case for another two?*

I.

From 1953:

[He] suddenly knew that he was tired. He always knew
when his body or his mind had had enough and he always
acted on the knowledge. This helped him to avoid stale-
ness and the sensual bluntness that breeds mistakes.

   He shifted himself unobtrusively away from the rou-
lette he had been playing and went to stand for a moment
at the brass rail which surrounded breast-high the top
table in the 'salle privée'.

ANSWERS ON PP. 132–4

2.

From 1969:

[He] was a man to whom everybody came for help, and never were they disappointed. He made no empty promises, nor the craven excuse that his hands were tied by more powerful forces in the world than himself. It was not necessary that he be your friend, it was not even important that you had no means with which to repay him. Only one thing was required. That you, *you yourself*, proclaim your friendship. And then, no matter how poor or powerless the supplicant, he would take that man's troubles to his heart . . . His reward? Friendship . . . And perhaps, to show respect only, never for profit, some humble gift – a gallon of homemade wine or a basket of peppered *tarelles* specially baked to grace his Christmas table. It was understood, it was mere good manners, to proclaim that you were in his debt and that he had the right to call upon you at any time to redeem your debt by some small service.

3.

From 1925:

I was still with Jordan Baker. We were sitting at a table with a man of about my age and a rowdy little girl who gave way upon the slightest provocation to uncontrollable laughter. I was enjoying myself now. I had taken two finger bowls of champagne and the scene had changed before my eyes into something significant, elemental and profound.

At a lull in the entertainment the man looked at me and smiled.

'Your face is familiar,' he said politely. 'Weren't you in the Third Division during the war?'

4.

*From 1847 – where a stranger has just
fallen off his horse:*

Q4

His figure was enveloped in a riding cloak, fur collared and steel clasped; its details were not apparent, but I traced the general points of middle height, and considerable breadth of chest. He had a dark face, with stern features and a heavy brow; his eyes and gathered eyebrows looked ireful and thwarted just now; he was past youth, but had not reached middle-age; perhaps he might be thirty-five. I felt no fear of him, and but little shyness . . .

'I cannot think of leaving you, sir, at so late an hour, in this solitary lane, till I see you are fit to mount your horse.' . . .

'I should think you ought to be at home yourself,' said he, 'if you have a home in this neighbourhood. Where do you come from?'

'From just below; and I am not at all afraid of being out late when it is moonlight . . .'

'You live just below – do you mean at that house with the battlements?' pointing to Thornfield Hall, on which the moon cast a hoary gleam.

ANSWERS ON PP. 132–4

ROUND 4

CONNECT THREE

Can you link the three items in each case?

1.

G. K. Chesterton's priest detective

The courageous Anna Fierling in a play by Bertolt Brecht

Ivan Petrovich Voynitsky in a play by Anton Chekhov

2.

Graham Greene's Scobie in Sierra Leone

Joseph Conrad's Kurtz in the Congo

William Boyd's Logan Montstuart in Uruguay, Britain, Spain, Portugal, the Bahamas, Switzerland, France, America and Nigeria

ANSWERS ON PP. 135–7

3.

The literary character who grows up in a Victorian work-house overseen by Mr Bumble the beadle

The Victorian socialist, poet and textile designer whose novels included *News from Nowhere*

The Puffin Classic by Noel Streatfeild subtitled *A Story of Three Children on the Stage*

Q4

4.

_____ S. Buck, the first American woman to win the Nobel Prize in Literature

_____ Wax, whose books on mental health include *A Mindfulness Guide for the Frazzled* and *How to be Human*

The _____, the Wilkie Collins book that T. S. Eliot considered the first British detective novel (as have many other people)

5.

A provincial lady (E. M. Delafield)

A nobody (George and Weedon Grossmith)

A wimpy kid (Jeff Kinney)

6.

James Hilton's horizon

John Milton's paradise

Marcel Proust's time

Q4

ANSWERS ON PP. 135–7

ROUND 5

QUICKFIRE: EDUCATED

Literature and learning . . .

Q4

1.

Which journalist and newspaper interviewer wrote *An Education*, a 2009 memoir of her teenage love affair with a much older man?

2.

Which fictional school, created by the cartoonist Ronald Searle, featured in seven films between 1954 and 2009?

3.

In a play by Christopher Marlowe, which scholar at the University of Wittenberg pledges his body and soul to the Devil in return for 24 years of being able to do whatever he likes?

122

ANSWERS ON PP. 138–9

4.

Who wrote the play *The School for Scandal*, first performed at the Drury Lane Theatre in 1777?

5.

Miss Cackle's Academy for Witches features in the children's classic *The Worst Witch* – by whom?

6.

George, Martha, Honey and Nick are the four characters in which play by Edward Albee, where the action takes place in the drunken aftermath of a university faculty party?

7.

Which 1959 novel by E. R. Braithwaite, based on his own experiences of being a black teacher in London, became a film of the same name starring Sidney Poitier and Lulu?

8.

Who won the 2006 Man Booker Prize with *The Inheritance of Loss*, in which one of the main characters has a relationship with her maths tutor?

9.

Which aristocratic detective, who graduated from Oxford with a first-class degree – while also becoming a cricket blue – was created by Dorothy L. Sayers?

10.

Which Nigerian former teacher wrote the 1958 novel *Things Fall Apart*, one of the first African novels to receive global acclaim?

Q4

11.

In a series of books by Rick Riordan that begins with *The Lightning Thief*, what's the name of the eponymous main character whose teacher Mrs Dodds turns out to be a mythological Fury?

12.

In 1965 a novel by John Williams about a university teacher in Missouri was published and sold fewer than 2,000 copies. It was republished in 2003, nine years after Williams's death, and became a worldwide bestseller. What was the novel's title – which was also the main character's surname?

ANSWERS
TO QUIZ 4

ROUND I

1.

Tina Brown

2.

Becky Thatcher

3.

East of Eden

4.

Melbourne. (Like Humphries himself, the book does move on from there. The title *More Please* comes from the first words that Humphries apparently ever spoke.)

5.

Cameron Post

6.

James Baldwin

7.

Angela Carter

8.

John Kennedy Toole

9.

Kate Bush (the single being 'Wuthering Heights')

10.

Gina Ford

11.

Samuel Johnson

12.

Truman Capote

Bonus:

The Will to Power is a book by Friedrich Nietzsche, published after he died and based on notes he had left behind. It was assembled by his anti-Semitic sister Elisabeth, who has been accused since of shaping the material to fit her own German nationalist ideas and so of being responsible for Nietzsche's appeal to the Nazis. When she died in 1935, Adolf Hitler was among the mourners at her funeral.

ROUND 2

1.

Jamie Oliver, and I'm willing to bet that almost everybody reading this now – and almost everybody they know – owns at least one of his books. The catchphrase is 'lovely jubbly', as also used by Del Boy in *Only Fools and Horses*.

2.

William Blake, author of the much-sung 'Jerusalem' (aka 'And Did Those Feet in Ancient Time'), whose funeral was attended by only five people. Robert's teaching, it seems, came to Blake in a vision. He also seems to have spotted his first angels at the age of nine, in Peckham Rye.

3.

Louisa May Alcott, whose family provided a safe house for runaway slaves when she was growing up, and who later campaigned for women's suffrage. Even more unusually for a woman of the time, she was also a keen runner.

4.

Leo Tolstoy, who in his radically Christian later life managed to turn himself, after an agonising series of personal struggles, into a non-smoking, non-drinking vegetarian admirer of peasants – although he didn't always stay on the smoking wagon. Sex, meanwhile, remained a problem. In a diary entry of 1909, Tolstoy lamented its continuing temptation and wished that it 'could be instilled into people in childhood and also when fully mature that the sexual act is a disgusting, animal act'. He was 80 at the time.

A 4

5.

John Updike, whose middle name was Hoyer because that was his mother's maiden name, and whose Rabbit books began with *Rabbit, Run* in 1960, continuing approximately every ten years from there. Rabbit is from Pennsylvania, like Updike – hence that interest in James Buchanan. *Couples*, a book about suburban adultery (like many of his others), remained his biggest selling novel.

6.

E. L. James. In fact, *Fifty Shades of Grey*, *Fifty Shades Darker* and *Fifty Shades Freed* were Britain's bestselling novels of 2012 by some margin – and the same year *Time* magazine named her as one of the 'World's 100 Most Influential People'. The *Fifty Shades* trilogy, she later reflected, 'was my midlife crisis, writ large'. She also wrote Britain's bestselling book of 2015: *Grey*, which told the story from Christian's point of view.

A 4

ROUND 3

I.

James Bond in *Casino Royale* by Ian Fleming. And with that, Bond duly collects his winnings and retires, pausing only to notice the villainous Le Chiffre winning millions of francs at that top table. If, at the end of 1953, you'd asked almost anybody in Britain what had been the year's most significant national events, it wouldn't have been hard to predict their replies: the Queen's Coronation and a British team conquering Everest. (Never mind that the two men who made it to the top were from New Zealand and Nepal.) Yet when it comes to Britain's global reach since then, a better answer – unimaginable at the time, admittedly – might have been the publication of this novel written by a 43-year-old bachelor to take his mind off 'the agony' of getting married.

2.

Don Vito Corleone in *The Godfather* by Mario Puzo – first seen on the wedding day of his daughter Connie. Francis Ford Coppola's 1972 film became the highest grossing of all time, until it was overtaken by 1976's *Jaws*. And in 2005 the American Film Institute voted Vito's 'I'm going to make him an offer he can't refuse' the second greatest quote in 100 years of US movie history. It was beaten only by Rhett Butler's 'Frankly my dear, I don't give a damn' from *Gone with the Wind*. (Another Marlon Brando character was at number three, thanks to *On the Waterfront*'s 'You don't understand! I coulda had class. I coulda been a contender. I could've been somebody, instead of a bum, which is what I am.')

A4

3.

After hearing plenty about his glamorous new neighbour, the narrator Nick Carraway finally meets (Jay) Gatsby in this passage, more than a quarter of the way through *The Great Gatsby* by F. Scott Fitzgerald. The meeting takes place at one of Gatsby's celebrated parties – although only after they've talked about their shared experiences of the First World War does Nick discover who he's just met.

4.

Mr Rochester in Charlotte Brontë's *Jane Eyre*. By this stage Jane is living in Mr R's Thornfield Hall as a governess – but, while she's heard plenty about him, she's never seen the man himself, who's been away since she arrived. Only after she's helped him back on to his horse and walked back to Thornfield does Jane discover who she's just met.

A 4

ROUND 4

I.

Family relatives: Chesterton's <u>Father</u> Brown; Brecht's <u>Mother</u> Courage in *Mother Courage and Her Children*; Chekhov's <u>Uncle</u> Vanya.

2.

Main characters in novels with 'Heart' in the title: Greene's *The <u>Heart</u> of the Matter*; Conrad's *<u>Heart</u> of Darkness*; Boyd's *Any Human <u>Heart</u>*.

3.

Dances: Dickens's Oliver <u>Twist</u>; William <u>Morris</u>; *<u>Ballet</u> Shoes*.

4.

Gemstones: <u>Pearl</u> S. Buck; <u>Ruby</u> Wax; *The <u>Moonstone</u>* – according to Eliot, 'the first, the longest, and the best of modern English detective novels'. (And just in case you're not up to speed on Pearl S. Buck, her work may be less read these days but back in the 1930s she achieved the writers' dream of combining enormous sales with enormous critical acclaim. Her book *The Good Earth* – based on her experiences as the daughter of US missionaries in China – was America's bestselling novel of both 1931 and 1932. It also won the 1932 Pulitzer Prize for Fiction – and she was awarded the Nobel Prize six years later.)

A 4

5.

Fictional diarists: Delafield's comic and heavily autobiographical *The Diary of a Provincial Lady* was a bestseller when it came out in 1930 and has never been out of print since. (Jilly Cooper is a particular fan.) The Grossmiths' *The Diary of a Nobody* is the book that gave us Charles Pooter, a good-hearted suburbanite whose surname has now come to mean someone particularly conventional and unimaginative – which seems a bit harsh to me. Jeff Kinney's *Diary of a Wimpy Kid* series has been raking in the sales since the first book was published in 2007, with Kinney playing cameo roles in the film versions.

6.

Lost things: Hilton's <u>Lost</u> Horizon was the book that first brought us the utopian Himalayan kingdom of Shangri-la – not, as you may think, an ancient legend but something thought up by one British author in 1933. (Hilton's other enduring work is the novella *Goodbye, Mr Chips*.) Milton's *Paradise <u>Lost</u>* is an epic about the fall and redemption of humankind, of which Dr Johnson once said that no reader 'ever wished it longer than it is'. And Proust's *In Search of <u>Lost</u> Time* isn't terribly short either. In fact, it's named as the longest novel ever by *Guinness World Records*.

A 4

ROUND 5

1.

Lynn Barber (the film version, written by Nick Hornby, was based on a shorter version of the same story that Barber had earlier published in *Granta* magazine)

2.

St Trinian's

3.

Dr Faustus (in *The Tragical History of the Life and Death of Doctor Faustus*)

4.

Richard Brinsley Sheridan

5.

Jill Murphy

6.

Who's Afraid of Virginia Woolf?

7.

To Sir, With Love

8.

Kiran Desai

A 4

9.

Lord Peter Wimsey

10.

Chinua Achebe

11.

Percy Jackson

12.

Stoner

Bonus:

Educated is a 2018 memoir by Tara Westover about her journey from a fundamentalist Mormon background in rural
Idaho to completing a doctorate at Cambridge University.

QUIZ 5

ROUND I

QUICKFIRE:
THE MUSIC SHOP

Books and music . . .

1.

Q5

The Scottish indie band Josef K took their name from a character created by whom?

2.

The first part of T. H. White's Arthurian work *The Once and Future King* is *The Sword in the Stone*. What's the fourth – which (apart from having a '*The*' at the beginning) shares its title with the biggest selling single of the twentieth century?

3.

Which Haruki Murakami novel shares its title with that of a Beatles song?

4.

Who's the only writer whose full name is mentioned in a Beatles song?

5.

The 2018 book *Rise Up* is by which British rapper, whose real name is Michael Ebenazer Kwadjo Omari Owuo Jr?

6.

The film *What's Love Got to Do With It?* was based on whose autobiography?

Q5

7.

Who wrote the twelve-novel series *A Dance to the Music of Time*?

8.

According to Kurt Cobain, Nirvana's song 'Scentless Apprentice' was based on which Patrick Süskind novel?

9.

According to Mick Jagger, the Rolling Stones' 'Sympathy for the Devil' was partly inspired by which Mikhail Bulgakov novel?

ANSWERS ON PP. 163–4

10.

Which dramatist and songwriter wrote the plays *Blithe Spirit* and *Hay Fever*?

11.

Set the Boy Free was the autobiography of which musician, whose bands have included The Cribs, Electronic and The Smiths?

12.

What's the name of the band in the title of a 2019 novel by Taylor Jenkins Reid?

Q5

145

ROUND 2

NAME THE AUTHOR

*Can you guess the writer from these clues
(and, of course, the fewer you need the better)?*

1.

A.   He was born in Bloemfontein – then a town in the Orange Free State, now a town in South Africa.

B.   His first book was *A Middle English Vocabulary* (1922).

C.   His most famous work was voted the greatest book of the twentieth century in a Waterstones poll – and Britain's favourite ever novel in the BBC's 'Big Read' project of 2003.

D.   That work begins on Bilbo Baggins's eleventy-first birthday.

ANSWERS ON PP. 165–7

2.

A. In that same Waterstones poll, one of his novels came second and another was third.

B. One of his earlier novels was called *A Clergyman's Daughter*.

C. He took his pen name from a river in Suffolk . . .

D. . . . and his real name was Eric Blair.

3.

A. She was made a Dame of the British Empire in 1999 for services to literature.

B. Her first name is the same as that of Harold Pinter's writer wife.

C. Her sister is the novelist Margaret Drabble.

D. Her novels include *The Children's Book*, *The Virgin in the Garden* and the Booker Prize-winning *Possession*.

Q5

147

4.

A. He testified for the literary worth of *Lady Chatterley's Lover* when Penguin was prosecuted for obscenity after publishing an unexpurgated version of the novel.

B. He published his last novel in 1924, 46 years before his death.

C. He was known to friends by his middle name of Morgan.

D. Screen versions of his novels include David Lean's last film – and three Merchant Ivory adaptations.

Q5

5.

A. Anthony Hopkins won a BAFTA for playing him in a film of 1993.

B. His scholarly books include *The Allegory of Love: A Study in Medieval Tradition*.

C. His most famous books feature four siblings with the surname Pevensie.

D. His first names were Clive Staples.

ANSWERS ON PP. 165–7

6.

A. In 1946 she was refused a visa for America because she'd been a member of the Communist Party of Great Britain . . .

B. . . . although she was born in Dublin.

C. Between 1948 and 1963 she taught philosophy at Oxford, an interest reflected in many of her novels.

D. Kate Winslet and Judi Dench both played her in the same film, which depicted her succumbing to Alzheimer's disease.

Q5

ANSWERS ON PP. 165–7

ROUND 3

EXTRACTS: TV TIME

These extracts are all from books that have inspired celebrated television drama series – the first one in the 1980s, the rest in the 2010s. Can you name the novel in each case for one point, and the author for another?

I.

From a novel of 1945:

I knew Sebastian by sight long before I met him. That was unavoidable for, from his first week, he was the most conspicuous man of his year by reason of his beauty, which was arresting, and his eccentricities of behaviour, which seemed to know no bounds. My first sight of him was in the door of Germer's, and, on that occasion, I was struck less by his looks than by the fact that he was carrying a large teddy-bear.

2.

From an autobiography of 1956:

Scarcely had we settled into the strawberry-pink villa before Mother decided that I was running wild, and that it was necessary for me to have some sort of education. But where to find this on a remote Greek island? As usual when a problem arose, the entire family flung itself with enthusiasm into the task of solving it. Each member had his or her own idea of what was best for me, and each argued with such fervour that any discussion about my future generally resulted in an uproar . . .

'He appears to have only one interest,' said Larry bitterly, 'and that's this awful urge to fill things with animal life. I don't think he ought to be encouraged in *that*. Life is fraught with danger as it is. I went to light a cigarette only this morning and a damn' great bumble-bee flew out of the box.'

'It was a grasshopper with me,' said Leslie gloomily.

3.

From a non-fiction book of 2016:

After he had been cautioned, [he] replied, 'I hear what you say. I am totally innocent of this charge and will vigorously challenge it.' He was then offered coffee and sandwiches before being taken to Minehead Magistrates' Court a few hundred yards away. In the courtroom [he] was charged with conspiracy to murder as well as incitement to murder – the most serious charges ever levelled against a sitting Member of Parliament.

ANSWERS ON PP. 168–9

4.

From a novel of 1985:

The lawns are tidy, the façades are gracious, in good repair; they're like the beautiful pictures they used to print in the magazines about homes and gardens and interior decoration. There is the same absence of people, the same air of being asleep. The street is almost like a museum, or a street in a model town constructed to show the way people used to live. As in those pictures, those museums, those model towns, there are no children.

This is the heart of Gilead, where the war cannot intrude except on television.

Q5

ROUND 4

ODD ONE OUT

In each of these batches of four,
what item doesn't belong and why?

I.

A. Anna Sewell's only novel

B. The first Australian to win the Nobel Prize in Literature

C. James Ellroy's Dahlia

D. The singer and TV presenter whose 2003 autobiography was called *What's It All About?* – the first four words of one of her most famous songs

ANSWERS ON PP. 170–72

ROUND 4

2.

A. The first volume of Maya Angelou's autobiography

B. The first novel by S. J. Watson, which became both a bestseller and a film starring Nicole Kidman

C. Robert Graves's fictional autobiography of the fourth Roman emperor

D. The Nicci French novel that shares a title with the first British number one single by The Fugees

3.

Q5

A. Ford Madox Ford's quartet of novels about the First World War, which was turned into a 2012 TV series written by Tom Stoppard and starring Benedict Cumberbatch

B. The three words that complete this quotation from a 1992 book by Francis Fukuyama: 'What we may be witnessing is not just the end of the Cold War, or the passing of a particular period of post-war history, but the . . .'

C. The first of Shakespeare's plays alphabetically

D. Karen Armstrong's 1996 work about the Book of Genesis that takes its title from the first three words of Genesis

4.

A. George Gissing's 1891 novel about literary London

B. The 1932 Aldous Huxley novel that's set in AD 2540

C. The city where Tom Wolfe's *The Bonfire of the Vanities* is set

D. The 1980 children's book by Prince Charles

5.

A. The Ernest Hemingway novel featuring Santiago the Fisherman

B. The H. G. Wells novel about a mad scientist who creates half-human creatures by practising vivisection on animals

C. The 1883 novel that features a voyage aboard the *Hispaniola*

D. Bill Bryson's 1995 travelogue about Britain

Q5

6.

A.   Anna Bouverie in a 1992 novel by Joanna Trollope

B.   Clare Abshire in a 2003 novel by Audrey Niffenegger

C.   Sir Robert Chiltern in the title of a play by Oscar Wilde

D.   A five-times-married pilgrim in Chaucer's *Canterbury Tales*

ANSWERS ON PP. 170–72

ROUND 5

QUICKFIRE:
THINK OF A NUMBER

All the answers here contain a number . . .

1.

Which 1853 memoir by Solomon Northup became a best-seller 160 years later thanks to an Oscar-winning film version?

2.

What number are the clocks striking in the first sentence of George Orwell's *Nineteen Eighty-Four*?

3.

Which Gabriel García Márquez novel begins with Colonel Aureliano Buendía facing the firing squad?

4.

Richard Hannay is the main character of which 1915 novel by John Buchan?

5.

Something Happened, *Good as Gold* and *God Knows* were this author's second, third and fourth novels – what was his first?

6.

What do you get if you add the number of Bennet sisters in Jane Austen's *Pride and Prejudice* to the number of Kara-mazov brothers in Fyodor Dostoevsky's *The Brothers Karamazov*?

7.

And what do you get if you add the number of Beatrix Potter's bad mice to the number of Julian Barnes's chapters in the history of the world?

8.

What was the first novel by Flann O'Brien?

9.

In which Kurt Vonnegut novel is Billy Pilgrim a prisoner of war in Dresden during the Allied bombing of the city?

10.

Which play by Reginald Rose became a 1957 film starring Henry Fonda as a heroic campaigner for justice?

11.

In which E. Nesbit novel do a group of siblings find a bad-tempered sand-fairy called the Psammead?

Q5

12.

What's the main title of the global bestseller of 2018 by the Canadian psychologist Jordan B. Peterson, subtitled *An Antidote to Chaos*?

ANSWERS
TO QUIZ 5

ROUND 1

1.

Franz Kafka (Josef K appears in *The Trial* and the short story 'A Dream')

2.

The Candle in the Wind

3.

Norwegian Wood

4.

Edgar Allan Poe – 'Man, you should have seen them kicking Edgar Allan Poe' from 'I Am the Walrus'. (Poe is also on the cover of *Sgt. Pepper's Lonely Hearts Club Band*.)

5.

Stormzy

6.

Tina Turner

7.

Anthony Powell

8.

Perfume

9.

The Master and Margarita

10.

Noël Coward

11.

Johnny Marr

12.

Daisy Jones and the Six

Bonus:

The Music Shop is a 2017 novel by Rachel Joyce about a 1980s record shop whose owner is holding out against the rise of CDs.

ROUND 2

1.

J. R. R. Tolkien, who moved to England with his family when he was three – and later fought in the Battle of the Somme. After being demobilised, he became (and remained) an academic, specialising in Anglo-Saxon and medieval language and literature – and that most famous work is, of course, *The Lord of the Rings*.

A5

2.

George Orwell, with *Nineteen Eighty-Four* and *Animal Farm* at two and three respectively in the Waterstones poll. For the record, *Nineteen Eighty-Four* was voted in at eight in the BBC's 'Big Read' and *Animal Farm* a surprisingly low 46.

3.

A. S. Byatt. The A and S stand for Antonia (as in Pinter's wife Antonia Fraser) and Susan. Drabble was also her maiden name, and her supposedly troubled relationship with Margaret has been a source of recurring newspaper speculation/delight. According to Margaret, though, it's 'normal sibling rivalry', and according to Antonia 'terribly overstated by gossip columnists'.

4.

E. M. Forster, whose first names were Edward Morgan. His last novel, and Lean's last film, was *A Passage to India* – after which, he said in a 1958 BBC documentary, 'somehow I dried up'. Those Merchant Ivory adaptations were of *A Room with a View*, *Howards End* and *Maurice*, a novel about a gay relationship that Forster thought too controversial to publish – mainly, it seems, because it had a happy ending (although he may also have been worried about outing himself). In the event, it appeared in 1971, the year after his death.

5.

C. S. Lewis, who was played by Hopkins in *Shadowlands*: a film about his relationship with an American woman called Joy Davidman, which came in later life after he'd spent most of his adulthood as an archetypal bachelor don at Oxford. Pevensie is the surname of Peter, Susan, Edmund and Lucy in *The Chronicles of Narnia*.

6.

Iris Murdoch, who joined the Communist party while a student at Oxford in 1938 and left in 1942. (She was later allowed into America.) The film was *Iris*, based on the memoir by her husband John Bayley – himself an Oxford professor of English.

I.

Brideshead Revisited by Evelyn Waugh – the first time that the narrator Charles Ryder sees Sebastian Flyte in 1920s Oxford. (Germer's was a famous real-life hairdresser's in the city of the time.) Thanks not least to the hugely successful 1980s TV version, starring Jeremy Irons as Charles, *Brideshead* – published in 1945 – is perhaps Waugh's most famous novel now. But he himself was a little embarrassed by it, and in 1959 brought out a revised edition that toned down some of the more purple passages. It also included a curiously apologetic preface in which Waugh explained that he had written the book during the Second World War in 'a bleak period of present privation and threatening disaster' – which is why it was 'infused with a kind of gluttony, for food and wine, for the splendours of the recent past, and for rhetorical and ornamental language, which now with a full stomach I find distasteful'. Or, in more simple terms: 'I piled it on rather.'

2.

My Family and Other Animals by Gerald Durrell, the at times quite loose basis of the series *The Durrells*. (There was also a one-off BBC series in 1987.) Critics have pointed out some factual inaccuracies in Durrell's account of his childhood in Corfu – but his writer brother Lawrence/Larry clearly felt that, in essence, it was disturbingly trustworthy. 'This is a very wicked, very funny, and I'm afraid rather truthful book,' he once said, 'the best argument I know for keeping thirteen-year-olds at boarding-schools and not letting them hang about the house listening in to conversations of their elders and betters.'

3.

A Very English Scandal by John Preston, the story of how the popular Liberal Party leader Jeremy Thorpe (whose name I cunningly omitted from the extract) came to be tried for the attempted murder of his former lover Norman Scott – and, even more shockingly, acquitted.

A5

4.

The Handmaid's Tale by Margaret Atwood, set in the religious totalitarian Republic of Gilead in a near-future New England. As well as the TV series, the novel has been turned into an opera, a ballet, a BBC radio series, a one-woman stage show, several theatre versions, a concept album by the band Lakes of Canada and a 1980 film with a screenplay by Harold Pinter.

ROUND 4

1.

The odd one out is **B**, Patrick <u>White</u> – who won the Nobel Prize in 1973 – because the others are all blacks. Anna Sewell's <u>Black</u> Beauty, narrated by the horse of the title, was written the year before she died, aged 58. James Ellroy's The <u>Black</u> Dahlia was based on a real-life murder case from the 1940s that was particularly resonant for Ellroy as his own mother had been murdered when he was ten. The singer and TV presenter is Cilla <u>Black</u> – whose 1966 hit 'Alfie' begins 'What's it all about, Alfie?'

2.

The odd one out is **D**, *Killing <u>Me</u> Softly* – because all the others have titles containing 'I': *<u>I</u> Know Why the Caged Bird Sings* by Maya Angelou; Watson's *Before <u>I</u> Go to Sleep*; and Graves's *<u>I</u>, Claudius*. Despite the acclaim they received – and the sales they notched up – Graves was always quite sniffy about *I, Claudius* and its sequel *Claudius the God*, both of which he wrote in eight months purely, he

A5

170

claimed, for the cash. 'Neither of them is of any real worth,' he once said – although, on the plus side, his cash-raising plan did work: 'Claudius has been very helpful in the money way. I am now able to support my children.'

3.

The odd one out is **D**, *In the Beginning* (as in 'In the beginning, God created the heavens and the earth'). The other three are all ends: the Ford quartet is collectively known as *Parade's End*; Fukuyama prematurely predicted the 'end of history' in the book *The End of History and the Last Man*; and the first Shakespeare play alphabetically is *All's Well That Ends Well*.

4.

The odd one out is **D**, *The Old Man of Lochnagar* – because the others are a lot newer: Gissing's *New Grub Street*; Huxley's *Brave New World*; and Wolfe's *New* York during the 1980s, when Wall Street traders considered themselves 'masters of the universe'.

5.

The odd one out is **A**, *The Old Man and the Sea*, because the rest are all islands of various kinds. First came Wells's *The Island of Doctor Moreau*, published in 1896, when both vivisection and Darwinian evolution were hot topics, then *Treasure Island* by Robert Louis Stevenson and finally Bryson's *Notes from a Small Island*.

6.

The odd one out is **C**, because Sir Robert Chiltern is the person referred to in the title of Wilde's *An Ideal Husband* – and the others, by contrast, were all wives. Anna Bouverie is the eponymous main character of Trollope's *The Rector's Wife*, and Clare Abshire of Niffenegger's *The Time Traveler's Wife*. The Wife of Bath is the only Chaucer pilgrim whose prologue to the tale she tells is longer than the tale itself – and she takes the chance to defend her lusty life, to boast about the control she's had over her husbands and to explain that her preference is for rich and elderly ones.

A5

ROUND 5

1.

Twelve Years a Slave

2.

Thirteen ('It was a bright cold day in April, and the clocks were striking thirteen.')

3.

One Hundred Years of Solitude

4.

The Thirty-Nine Steps

5.

Catch-22 – by Joseph Heller

6.

Eight (five sisters, three brothers)

7.

12½ (two bad mice, 10½ chapters)

8.

At Swim-Two-Birds

9.

Slaughterhouse-Five

10.

Twelve Angry Men

11.

Five Children and It – the Psammead being the 'It'

12.

12 Rules for Life

Bonus:

Think of a Number is a 1979 maths book for children by Johnny Ball that accompanied a BBC TV series of the same name. (Slightly confusingly, he published a different book with the same title in 2005.)

QUIZ 6

ROUND I

QUICKFIRE:
CAKES AND ALE

Tuck into a round on food and drink . . .

I.

Which novel of 1886 begins with the main character drunkenly selling his wife and baby to a passing sailor?

2.

In Roald Dahl's *Charlie and the Chocolate Factory*, what is Charlie's surname?

3.

Whose bestselling books on Asian cooking include *Complete Chinese Cookbook* and *Travels with a Hot Wok*?

4.

Wilbur Larch – played by Michael Caine in the film version – is the main character in which John Irving novel?

177

5.

Which literary character is first seen in his house at 7, Eccles Street, Dublin, preparing breakfast for his wife Molly?

6.

Which literary character successively eats one piece of chocolate cake, one ice-cream cone, one pickle, one slice of Swiss cheese, one slice of salami, one lollipop, one piece of cherry pie, one sausage, one cupcake, and one slice of watermelon?

7.

What was the title of Jeanette Winterson's first novel?

8.

Which alcoholic drink is the single-word pen name of the short-story writer whose real name was Hector Hugh Munro?

9.

Which (not very) alcoholic drink appears in the title of an eighteenth-century novel by Laurence Sterne?

10.

Which Martin Amis novel is narrated by the heavy-drinking John Self?

ANSWERS ON PP. 197–8

11.

Whose novels include *Peaches for Monsieur le Curé*, *The Lollipop Shoes*, *Blackberry Wine* and *Chocolat*?

12.

In the title of a 1983 children's book by Lynley Dodd, what's the name of the dog from Donaldson's Dairy?

Q6

179

ROUND 2

NAME THE AUTHOR

*Can you guess the writer from these clues
(and, of course, the fewer you need the better)?*

I.

A. As a cricketer, the only first-class wicket he took
was that of W. G. Grace – and although not known
as a poet, he did write a long poem about that.

B. One of his novels has the same title as Michael
Crichton's sequel to *Jurassic Park*.

C. His last major book was a history of spiritualism,
in which he was a firm believer.

D. In 1990 a museum dedicated to his most famous
character opened in Baker Street, London, bearing
the door number 221B.

2.

A.   She was Britain's bestselling novelist of the 1980s.

B.   Her first and second names begin with consecutive letters of the alphabet.

C.   Her last novel was *The Woman Who Went to Bed for a Year*.

D.   Her most famous character was a diarist with a long-standing crush on Pandora Braithwaite.

3.

A.   She was born in the same year as James Joyce was – and died in the same year that he did.

B.   She belonged to an artistic group that, according to Dorothy Parker, 'lived in squares, painted in circles and loved in triangles'.

Q6

C.   Her sister was the artist Vanessa Bell.

D.   One of her best-known novels concerns a day in the life of a woman called Clarissa Dalloway.

ANSWERS ON PP. 199–201

4.

A. He wrote the screenplay for what, in 1999, the British Film Institute named as the best British film of the twentieth century.

B. While working as a film critic himself, he was sued by Twentieth Century Fox for suggesting that the child star Shirley Temple was deliberately coquettish and that her appeal to some men might be sexual.

C. He wrote a sympathetic introduction to the autobiography by the Soviet spy Kim Philby, who by then had fled to Russia.

D. His novels have a variety of globe-trotting settings – and place names in their titles include Geneva, Brighton and Havana.

Q6

5.

A. His early novels were set in Ireland, where he lived for ten years, including during the Great Famine of the 1840s.

B. His mother was a writer, whose controversial 1832 book *Domestic Manners of the Americans* – based on her travels there – became a huge seller on both sides of the Atlantic.

C. While working for the Post Office, he introduced the pillar box to Britain.

D. He wrote a series of novels set in the fictional county of Barsetshire, and another featuring the Palliser family.

Q6

ANSWERS ON PP. 199–201

6.

A. He was the first person in history to win both the Nobel Prize in Literature and an Oscar.

B. In 1933, after a visit to Stalin's Soviet Union, he co-wrote a letter to the *Manchester Guardian* saying that 'everywhere we saw a hopeful and enthusiastic working-class . . . setting an example of industry and conduct which would greatly enrich us if our systems supplied our workers with any incentive to follow it'.

C. In 1914 excited (and accurate) rumours that the word 'bloody' was to be used in one of his plays caused the biggest traffic jam in London's West End since George V's Coronation of 1911.

D. That same play was later turned into the musical *My Fair Lady*.

Q6

ANSWERS ON PP. 199–201

ROUND 3

EXTRACTS: LIGHTING UP

Allen Lane, the founder of Penguin, famously wanted to publish books that cost the same as a packet of cigarettes. (These days, in fact – for rather different reasons – most paperbacks are a lot cheaper than that.) But can you identify these books where the characters have opted for the fags, as well, of course, as their authors? (The usual one point for the book and a second for the author.)

Q6

I.

From a novel of 1954:

A minute later Dixon was sitting listening to a sound like the ringing of a cracked door-bell as Welch pulled at the starter. This died away into a treble humming that seemed to involve every component of the car. Welch tried again; this time the effect was of beer-bottles jerkily belaboured. Before Dixon could do more than close his eyes he was pressed firmly back against the seat, and his

185

cigarette, still burning, was cuffed out of his hand into some interstice of the floor. With a tearing of gravel under the wheels the car burst from a standstill towards the grass verge, which Welch ran over briefly before turning down the drive. They moved towards the road at walking pace, the engine maintaining a loud lowing sound . . .

2.

From a non-fiction book of 1933 (it might help to know that a spike was a casual ward where vagrants could stay the night – and that the Tramp Major was the official who supervised it):

Once again it was jolly autumn weather, and the road was quiet, with few cars passing. The air was like sweet-briar after the spike's mingled stenches of sweat, soap, and drains. We two seemed the only tramps on the road. Then I heard a hurried step behind us, and someone calling. It was little Scotty, the Glasgow tramp, who had run after us panting. He produced a rusty tin from his pocket. He wore a friendly smile, like someone repaying an obligation.

'Here y'are, mate,' he said cordially. 'I owe you some fag ends. You stood me a smoke yesterday. The Tramp Major give me back my box of fag ends when we come out this morning. One good turn deserves another – here y'are.'

And he put four sodden, debauched, loathly cigarette ends into my hand.

ANSWERS ON PP. 202–4

3.

From a novel of 1993:

He found a tree that had not been damaged by shellfire and sat down beneath it, lighting a cigarette and sucking in the smoke. Before the war he had never touched tobacco; now it was his greatest comfort.

. . . The tunnellers had become more and more part of the army; although they had not undergone the humiliating drills and punishments handed out to the infantry before they were deemed ready to fight, they had lost the separate status they had had at first . . . They had become soldiers and were expected to kill the enemy not only by mining but with bayonet or bare hands if necessary.

4.

From a non-fiction book of 1992:

Q6

I remember the game for conventional reasons, for substitute Smith's late winner and thus a handy Cup win. But most of all I remember it as the only time in the 1980s and, hitherto, the 1990s, that I had no nicotine in my bloodstream for the entire ninety minutes . . . In October '89, after a visit to Allen Carr the anti-smoking guru, I went cold turkey for ten days, and this game came right in the middle of that unhappy period.

ANSWERS ON PP. 202–4

ROUND 4

ORDER, ORDER

*Can you place these three items in ascending,
correct or their usual order?*

I.

A. An anti-slavery work by Harriet Beecher Stowe
    that became America's bestselling novel of the
    nineteenth century

B. Jo Nesbø's detective

C. The former jockey whose novels include *Dead Cert*
    and *Dead Heat*

2.

A. The 1908 novel by G. K. Chesterton subtitled
    A Nightmare

B. The Ian McEwan novel set on the day of the march
    against the Iraq war

C. Robinson Crusoe's island servant

188

ANSWERS ON PP. 205-7

3.

A. Edward de Waal's hare's eyes

B. Anne Shirley in the books by L. M. Montgomery

C. The Brothers Grimm tale whose German title is *Rotkäppchen*

4.

A. Ford Madox Ford's soldier, according to the title of a 1915 novel

B. How Nora Ephron felt about her neck in a 2007 collection of essays

C. The Hans Christian Andersen story whose Danish title is *Den grimme ælling*

Q6

5.

A. *Coming Up for* ___ (a 1939 novel by George Orwell)

B. ___ *Milk* (a 2016 novel by Deborah Levy)

C. *The Moon's a* _____ (a 1971 memoir by David Niven)

189

6.

A. Flann O'Brien's policeman in a posthumously published novel

B. Simone de Beauvoir's 1949 work that became a cornerstone of the feminist movement in the 1960s and 1970s

C. Ivan Turgenev's love, according to the title of an 1860 novella

ROUND 5

QUICKFIRE:
THIS SPORTING LIFE

Books and sport . . .

I.

Bare-knuckle fighting plays a central role in which 1996 novel by Chuck Palahniuk?

2.

Table tennis plays a central role in whose 1999 novel *The Mighty Walzer*?

3.

Ireland's unexpected progress in the 1990 football World Cup is the background to whose novel *The Van*?

4.

Whose second novel was *Human Croquet* – the follow-up to *Behind the Scenes at the Museum*?

5.

Which Charles Dickens novel, the first he published, features a cricket match between All Muggleton and Dingley Dell?

6.

No Spin was the 2018 autobiography by which Australian cricketer?

7.

Believe was the 2017 autobiography by which British Olympian – the first woman ever to win an Olympic gold medal for boxing?

8.

The author of the autobiography *I Am Zlatan Ibrahimović* (clue: his name is Zlatan Ibrahimović) played football for which international side?

9.

Who boasted of the time he swam the Hellespont, between Europe and Asia, in his long poem *Don Juan*?

10.

Who won the William Hill Sports Book of the Year in 2000 with the autobiography *It's Not About the Bike: My Journey Back to Life*?

11.

Which eponymous Shakespeare character is insulted by a gift of tennis balls from the Dauphin of France?

Q6

12.

Whose books on mountaineering include *The Beckoning Silence*, *Dark Shadows Falling* and *Touching the Void*?

ANSWERS
TO QUIZ 6

ROUND I

1.

The Mayor of Casterbridge – by Thomas Hardy

2.

Bucket

3.

Ken Hom

4.

The Cider House Rules

5.

Leopold Bloom – in James Joyce's *Ulysses*

6.

The Very Hungry Caterpillar – in the book by Eric Carle

7.

Oranges Are Not the Only Fruit

8.

Saki

9.

Shandy (*Tristram Shandy* – or more precisely/pedantically *The Life and Opinions of Tristram Shandy, Gentleman*)

10.

Money

11.

Joanne Harris

12.

Hairy Maclary

Bonus:

Cakes and Ale is a 1930 novel about literary London by Somerset Maugham. Its title is taken from Sir Toby Belch's question to the puritanical Malvolio in Shakespeare's *Twelfth Night*: 'Dost thou think, because thou art virtuous, there shall be no more cakes and ale?'

ROUND 2

1.

Sir Arthur Conan Doyle, whose other claim to sporting fame is that, while staying in Davos in 1893, he introduced skiing to Switzerland. The book of his that shares a title with Michael Crichton is *The Lost World*, the first of three novels he wrote featuring Professor Challenger. Like Crichton's, it also features dinosaurs – although in Doyle's book, they've survived to the present day on a plateau in South America. Oh yes, and that most famous character of his was Sherlock Holmes – who lived at 221B Baker Street. (Oddly enough, despite its door number, the museum is situated between 237 and 241 Baker Street.)

A6

2.

Sue Townsend, the creator of Adrian Mole, who first appeared aged 13¾ in 1982 and was last seen approaching 40 in 2009's *Adrian Mole: The Prostrate Years*, where his love for Pandora remains both undimmed and unrequited. (Back in the first book, he'd sent her a poem for Valentine's Day that read: 'Pandora!/ I adore ya./ I implore ye/ Don't ignore me.')

3.

Virginia Woolf (1882–1941), who along with Vanessa was part of the Bloomsbury Group. In true triangular fashion, Vanessa had two sons by her art critic husband Clive Bell and a daughter by artist Duncan Grant, who also had relationships with Lytton Strachey and John Maynard Keynes (among other men). The daughter, Angelica, was brought up as Clive's child and didn't find out her real father's identity until she was eighteen. Virginia, meanwhile, had several love affairs with women during her own marriage to Leonard Woolf – most famously with Vita Sackville-West.

4.

A6

Graham Greene – author of *Dr Fischer of Geneva*, *Brighton Rock* and *Our Man in Havana*. His other fictional settings include Congo (*A Burnt-Out Case*), Vietnam (*The Quiet American*) and Haiti (*The Comedians*). Greene and Philby had worked together in the secret service and continued to write to each other after Philby fled to Moscow. The BFI-acclaimed film was *The Third Man* (1949), directed by Carol Reed.

5.

Anthony Trollope, who despite his success as a novelist continued to work for the Post Office until his early fifties. His mother Frances's book took a pretty dim view of Americans and she was particularly disgusted by slavery (and tobacco chewing). She concluded that: 'A single word indicative of doubt, that any thing, or every thing, in that country is not the very best in the world, produces an effect which must be seen and felt to be understood.'

6.

George Bernard Shaw, who won the Nobel Prize in 1925 and the Oscar in 1939 for his adaptation of his own play *Pygmalion* – the same one that caused that 1914 traffic jam. He was the only person to have won both awards until he was joined by Bob Dylan in 2016. (Shaw, incidentally, was not only a socialist and a vegetarian but also an enthusiastic campaigner for spelling reform. After the atom bombs were dropped on Japan, he wrote to *The Times* to protest – about the unnecessary 'b' on the end of the word 'bomb'.)

A6

ROUND 3

I.

Lucky Jim by Kingsley Amis, with the junior academic and main character Jim Dixon unwisely accepting a lift from his boss Professor Welch. *Lucky Jim* was Amis's first novel and such an immediate success that it enabled him to buy a television and have drink *in the house* – which his friend Philip Larkin regarded as the true test of affluence. Christopher Hitchens thought *Lucky Jim* the funniest book of the second half of the twentieth century, although not everyone was so keen at the time. In the *Times Literary Supplement*, the critic J. G. Weightman complained that the book had spread the impression that redbrick universities were 'peopled by beer-drinking scholarship louts, who wouldn't know a napkin from a chimney-piece and whose one ideal is to end their sex starvation in the arms of a big breasted blonde'.

2.

Down and Out in Paris and London by George Orwell. For this, his first book, the Old Etonian and former colonial policeman lived a life of poverty in both cities – and in the second part he also headed out of London to experience life on the road with his fellow 'tramps'. Or, as the writer V. S. Pritchett memorably put it, he 'went native in his own country'.

3.

Birdsong by Sebastian Faulks. The character is Jack Firebrace, whose job was to tunnel beneath German trenches so as to listen in, and to plant mines. Faulks's breakthrough novel was voted the nation's thirteenth favourite novel in the BBC's 2003 'Big Read' project. As for smoking during the First World War, supplying the troops with tobacco became a patriotic duty. There was a popular song called 'Don't Forget the Cigarettes for Tommy', and altogether the public subscribed enough money to provide British soldiers with 232,599,191 of them. (In the Second World War, the British government spent more on tobacco for the troops than on tanks, ships or planes.)

A 6

4.

Fever Pitch by Nick Hornby, his memoir of being an obses-sive Arsenal fan. 'Two down against Tottenham in a Cup semi-final at Wembley with eleven minutes to go and no fag? Inconceivable,' Hornby concludes – but he'd have to conceive it now that Wembley is a no-smoking stadium where you're not allowed to nip out for a cigarette and come back in either.

ROUND 4

1.

A, C, B – Tom, Dick and Harry:

Uncle <u>Tom</u>'s Cabin (whose saintly main character has since given rise to the insult 'Uncle Tom', for a black person who's overly subservient)

<u>Dick</u> Francis

<u>Harry</u> Hole

Sad added detail: At the age of 77, and suffering from dementia, Harriet Beecher Stowe started writing *Uncle Tom's Cabin* all over again.

2.

A, C, B – consecutive days of the week:

The Man Who Was <u>Thursday</u>

<u>Friday</u> (which is where the phrase 'Man Friday' – meaning a loyal personal assistant – came from, although the phrase is not used in the book)

<u>Saturday</u>

3.

C, A, B – red, amber, green:

Rotkäppchen is known in English as
Little <u>Red</u> Riding Hood

De Waal's *The Hare with <u>Amber</u> Eyes*, which won
the 2010 Costa Biography of the Year

Anne Shirley is better known as Anne
of <u>Green</u> Gables

4.

A, B, C – The Good, the Bad and the Ugly:

The <u>Good</u> Soldier

I Feel <u>Bad</u> About My Neck

The <u>Ugly</u> Duckling.

5.

B, A, C – 'hot-air balloon':

<u>Hot</u> Milk

Coming Up for <u>Air</u>

The Moon's a <u>Balloon</u>

A6

6.

C, B, A – first, second, third:

First Love

The Second Sex

The Third Policeman

ROUND 5

1.

Fight Club

2.

Howard Jacobson

3.

Roddy Doyle

4.

Kate Atkinson

5.

The Pickwick Papers

6.

Shane Warne

7.

Nicola Adams

8.

Sweden

9.

Lord Byron

10.

Lance Armstrong

11.

Henry V

A6

12.

Joe Simpson

Bonus:

This Sporting Life is a novel by David Storey in which the main character Arthur Machin tries to make it as a rugby league player (as Storey himself had been). It was made into a celebrated film in 1963, with Richard Harris as Machin, whose first name became Frank in the movie version.

QUIZ 7

ROUND 1

QUICKFIRE: FIRST PERSON

Twelve questions (or answers) featuring forenames . . .

1.

Who are the eponymous brother and sister in a picture book series by Lauren Child that became a TV series in 2005?

2.

Which eighteenth-century novel by Samuel Richardson is more than 1,500 pages long in the Penguin Classics edition?

Q7

3.

In the Wallander books by Henning Mankell, what is Wallander's first name?

ANSWERS ON PP. 231–2

4.

In the books by P. G. Wodehouse, what is Jeeves's first name?

5.

Anna of the Five Towns – set, like many of his later ones, in the Potteries area of Staffordshire – was an early novel by which writer?

6.

What girl's name is the title of an 1881 children's classic – and now one of the bestselling novels of all time – by the Swiss writer Johanna Spyri?

7.

What girl's name is the title of a children's classic by Eleanor H. Porter – and now, thanks to the book, a byword for someone who always looks on the bright side?

8.

John Boynton were the first two names of which Bradford-born novelist, critic and playwright, who died in 1984?

9.

Cecil Scott were the first two names of which writer, the creator of Horatio Hornblower, who serves in the Royal Navy during the Napoleonic Wars?

ANSWERS ON PP. 231–2

10.

Which forename is the title of a 1901 novel by Rudyard Kipling?

11.

Which William Styron novel became the film for which Meryl Streep won her first Best Actress Oscar, playing the character whose first name appears in the title?

12.

Whose first, third and fifth novels are *Amy and Isabelle*, *Olive Kitteridge* and *My Name Is Lucy Barton*?

Q7

ROUND 2

NAME THE AUTHOR

*Can you guess the writer from these clues
(and, of course, the fewer you need the better)?*

I.

A. After leaving college, he was a Christian missionary in Brazil.

B. He wrote America's bestselling novels of 1994, 1995, 1996, 1997, 1998, 1999, 2000, 2002, 2005, 2008 and 2011.

C. People who've starred in the film versions include Tom Cruise, Julia Roberts, Dustin Hoffman, Sandra Bullock, Kevin Spacey and Matt Damon . . .

D. . . . and all of them were playing lawyers.

Q7

ANSWERS ON PP. 233–5

NAME THE AUTHOR

2.

A. Although she's best known as a fiction-writer, her first five books – from *Double Persephone* to *The Animals in That Country* – were collections of poetry.

B. (One for crossword fans.) The copyright to her books is held by 'O. W. Toad'.

C. She was the first Canadian woman to win the Booker Prize . . .

D. . . . which she did in 2000 with *The Blind Assassin*, the novel of hers that succeeded *Alias Grace* and preceded *Oryx and Crake* – both of which were Booker shortlisted.

3.

A. During the 2017 General Election campaign, Jeremy Corbyn often quoted a stanza of his beginning, 'Rise, like lions after slumber/ In unvanquishable number!'

B. He was expelled from Oxford for writing an atheist pamphlet.

C. He drowned in 1822, shortly before his thirtieth birthday.

D. His poems include 'Adonaïs', 'Ode to the West Wind' and 'Ozymandias'.

Q7

4.

A. He described himself in *Who's Who* as 'poet and hack'.

B. He had a teddy bear called Archibald Ormsby-Gore.

C. He was the Poet Laureate between Cecil Day-Lewis and Ted Hughes.

D. He wrote the uncharacteristically unkind line 'Come friendly bombs and fall on Slough!'

5.

A. Before making it as a writer, he was a press officer for the Central Electricity Generating Board.

B. He was knighted in 2009 for services to literature.

C. The first book in his most famous series was *The Colour of Magic* and the 41st and final one *The Shepherd's Crown*.

D. The series was set on a planet called Discworld.

ANSWERS ON PP. 233–5

6.

A.   Her second novel popularised a female first name that up until then had been male.

B.   Her first novel was dedicated to William Makepeace Thackeray . . .

C.   . . . and had a heroine who rejected a marriage proposal from St John Rivers.

D.   She was the last of three famous sisters to die.

Q7

ROUND 3

EXTRACTS:
FAMOUS FIRST LINES . . .

. . . are of course a staple of quizzes about books. This, though, is a round about second lines. The quotations here are the words that come immediately after four of the most famous opening sentences in all fiction. Can you identify the book and author for a point each – and for a bonus point also say (or at least get pretty close to) what those celebrated first lines are?

I.

From an American novel of 1851:

Some years ago – never mind how long precisely – having little or no money in my purse, and nothing particular to interest me on shore, I thought I would sail about a little and see the watery part of the world.

ANSWERS ON PP. 236–7

2.

From a British novel of 1953:

When I came upon the diary it was lying at the bottom of a rather battered red cardboard collar-box, in which as a small boy I kept my Eton collars.

3.

From a British novel of 1813:

However little known the feelings or views of such a man may be on his first entering a neighbourhood, this truth is so well fixed in the minds of the surrounding families, that he is considered the rightful property of some one or other of their daughters.

4.

From a Russian novel of 1878:

Everything had gone wrong in the Oblonsky household.

Q7

221

ROUND 4

CONNECT THREE

Can you link the three items in each case?

1.

E. B. White's Wilbur

Hugh Lofting's Gub-Gub

Peppa in nearly 200 Ladybird books

2.

The author of 2019's *Where the Wild Cooks Go* – who was also the lead singer of the band Catatonia

The creator of the fictional pilot James Bigglesworth, better known as Biggles

'Mr Nice', according to the title of his autobiography, which chronicled his life of drug-smuggling

ANSWERS ON PP. 238–9

3.

The Jules Verne novel in which Professor Otto Lidenbrock begins the eponymous trip by descending into an Icelandic volcano

The bestselling 1992 self-help book by the American marriage counsellor John Gray, whose title has since become a well-known phrase

A 1995 novel by W. G. Sebald

4.

Penelope Lively's Booker Prize-winner

The only novel by Giuseppe Tomasi di Lampedusa

The character in a 1900 novel by L. Frank Baum who lacks courage

Q7

5.

The Günter Grass novel originally published in German as *Die Blechtrommel*

The children's classic by Ian Serraillier, published in America as *Escape from Warsaw*

The woman, born in 1947, who's now the bestselling American novelist of all time

6.

The only Shakespeare play with an animal in the title

The eighteenth-century author of *Joseph Andrews* and *Tom Jones*

The poet whose poems include 'Home-Thoughts, from Abroad' and 'The Pied Piper of Hamelin'

Q7

ANSWERS ON PP. 238–9

ROUND 5

QUICKFIRE:
LIVING WITH THE GODS

Books and religion . . .

1.

Whose first novel, set in a fourteenth-century Italian monastery, was *The Name of the Rose*?

2.

Which author, the first Mississippi-born winner of the Nobel Prize in Literature, wrote the books *Go Down, Moses* and *Absalom, Absalom!*?

3.

Which author, the first African-American winner of the Nobel Prize in Literature, wrote the books *Song of Solomon* and *Paradise*?

Q7

ANSWERS ON PP. 240–42

ROUND 5

4.

Which writer – the author of an enduring satire featuring a main character whose first name is Lemuel – became the Dean of St Patrick's cathedral in Dublin in 1713?

5.

The biblical phrase 'Moab is my washpot' provided the title for whose 1997 autobiography?

6.

What's the title of Richard Dawkins's bestselling 2006 book arguing the case for atheism?

7.

Who wrote the 1966 novel *The Fixer*, whose main character is a Jew living in Kiev, where he's arrested for the murder of a Christian boy during Passover?

8.

Which Swiss-German author wrote *Siddhartha*, whose main character seeks and finds enlightenment during the time of the Buddha?

9.

In which classic American novel by Nathaniel Hawthorne does Hester Prynne fall foul of her puritan community in seventeenth-century Massachusetts for having a baby while unmarried?

10.

As what is Fr Damien Karras known in the title of a 1970s bestseller by William Peter Blatty that became a celebrated film?

11.

A selection of sermons by which Christian minister, who died in 1968, is published under the title *A Gift of Love*?

12.

Whose first work of fiction was the collection *Scenes from Clerical Life*, published two years before her first novel *Adam Bede*?

Q7

ANSWERS
TO QUIZ 7

ROUND I

1.

Charlie and Lola

2.

Clarissa

3.

Kurt

4.

Reginald. (When Bertie Wooster finds out, he is stunned: 'It had never occurred to me before that he had a first name.')

5.

Arnold Bennett

6.

Heidi

7.

Pollyanna

8.

J. B. Priestley

9.

C. S. Forester. (In fact, he was born Cecil Louis Troughton Smith, before taking the pen name Cecil Scott Forester – and sticking to it so firmly that he even claimed in his autobiography that his parents were called Forester.)

10.

Kim

11.

Sophie's Choice

12.

Elizabeth Strout – in 1998, 2008 and 2016 respectively

Bonus:

First Person is a 2017 novel by Richard Flanagan, based on his own experience as a struggling young author of being asked to ghostwrite the memoirs of one of Australia's biggest ever con men. (And I suppose that you also get the bonus for <u>*The First Person*</u>: a short story collection by Ali Smith.)

ROUND 2

1.

John Grisham, who has been a Sunday school teacher for much of his adult life. While practising as a lawyer himself, he also served in the Mississippi House of Representatives, before the success of his second legal thriller, *The Firm*, enabled him to take up writing full-time. (The first was *A Time to Kill*, which then became a bestseller too.)

2.

Margaret Atwood, whose first novel was *The Edible Woman* in 1969. Before that, she was already highly respected for her poetry, and in 1966 won Canada's prestigious Governor General's Award for Poetry with *The Circle Game*. (And congratulations if you spotted that O. W. Toad is an anagram of Atwood.)

A7

3.

Percy Bysshe Shelley. The quotation used by Corbyn is from 'The Masque of Anarchy', written in response to the Peterloo Massacre; the stanza (and the poem) ends: 'Ye are many – they are few.' 'Adonaïs' was his elegy to John Keats, which Mick Jagger famously read from at the Rolling Stones concert in Hyde Park following the death of the band's former guitarist Brian Jones.

4.

John Betjeman, who kept his childhood bear all his life – along with a toy elephant, duly called Jumbo – and died with them both in his arms. That poem on Slough was a particular source of annoyance to Ricky Gervais's David Brent in the TV sitcom *The Office*, who was a proud resident of the town. 'I don't think you solve town planning problems by dropping bombs all over the place,' Brent once pointed out. 'He's embarrassed himself there.'

A7

5.

Terry Pratchett, author of the *Discworld* series, whose press officer job was based in the West Country and covered four nuclear power stations. 'By and large,' he later said, 'I worked for the nuclear industry during a period when we were telling the truth. Because they'd tried telling everything else and it hadn't worked.'

6.

Charlotte Brontë, who dedicated *Jane Eyre* to Thackeray because of her long-standing admiration for him. Rather awkwardly, she didn't know at the time that, like Mr Rochester in the novel, he had a wife with a serious mental illness. The novel that transformed a male name to a female one was *Shirley*, whose eponymous heroine was an independently minded heiress.

ROUND 3

I.

Moby-Dick by Herman Melville, the first line being 'Call me Ishmael.' In his acceptance speech for the 2016 Nobel Prize in Literature, Bob Dylan cited the novel, along with *All Quiet on the Western Front* and *The Odyssey*, as one of three big literary influences on his work – although it was later noted that many of his comments on Melville's book were suspiciously similar to those on the cribbing website SparkNotes.

2.

The Go-Between by L. P. Hartley, which begins 'The past is a foreign country: they do things differently there' before transporting us, and the elderly narrator Leo, back to his childhood in the long hot summer of 1900. Rather disappointingly, though, the phrase 'The past is a foreign country' wasn't the author's own work: it was first used by Hartley's good friend David Cecil in his inaugural lecture as Goldsmiths' Professor in 1949.

3.

Pride and Prejudice by Jane Austen – which means that the first line is (all together now): 'It is a truth universally acknowledged, that a single man in possession of a good fortune, must be in want of a wife.' As you've probably noticed, the first six words there have proved extremely useful for any journalist lacking a bit of inspiration who wants to start an article in a comic yet authoritative way. Rather disappointingly again, though, it may be that Austen wasn't being particularly original either: *Complete French Spelling Book* by Mr Gros, published in 1805, begins: 'It is a truth universally acknowledged, that our proficiency in any science, depends in a particular manner on first principles.'

4.

Anna Karenina by Leo Tolstoy – where everything had gone wrong in the household because Anna's brother Prince Oblonsky had been having an affair with the governess and his wife had recently found out. All of which allows Tolstoy to begin the novel with his much-quoted, if not necessarily true, observation: 'All happy families are alike but an unhappy family is unhappy after its own fashion.' That's the Penguin Classic version, but still full bonus points if you went for a similar translation of your own.

A7

ROUND 4

1.

Fictional pigs: Wilbur is in White's *Charlotte's Web*; Gub-Gub in Lofting's *Doctor Dolittle* series; and, as you might imagine, Peppa in Ladybird's Peppa Pig books.

2.

Plural versions of the writers of New Testament gospels: Cerys <u>Matthews</u> is the musician and food enthusiast; W. E. <u>Johns</u> created Biggles; and Mr Nice was Howard <u>Marks</u>. (Rather generously, the plural of male first names is also an acceptable link.)

A7

3.

Planets: Professor Otto Lidenbrock starts in Iceland in Verne's *Journey to the Centre of the <u>Earth</u>*; Gray's book is *Men Are from <u>Mars</u>, Women Are from <u>Venus</u>*; and Sebald's novel is *The Rings of <u>Saturn</u>*.

4.

Big cats: Lively's *Moon Tiger*; Tomasi's *The Leopard*, about a nobleman in Sicily watching his world disappear as Italy is united; Baum's Cowardly Lion – Baum being the author of *The Wonderful Wizard of Oz* on which the film *The Wizard of Oz* was based.

5.

Metals: Grass's *The Tin Drum*; Serraillier's *The Silver Sword*; and the record-holding Danielle Steel.

6.

Present participles: *The Taming of the Shrew*; Henry Fielding; Robert Browning. (And just for any fellow pedants reading this, I realise that 'taming' is acting as a noun in the Shakespeare title – but the word itself still counts as a present participle.)

A7

ROUND 5

1.

Umberto Eco

2.

William Faulkner. (*Absalom, Absalom!* also takes its title from a biblical character.)

3.

Toni Morrison (also the first black woman of any nationality to win the Nobel)

4.

Jonathan Swift – Lemuel is Gulliver's first name in *Gulliver's Travels*

5.

Stephen Fry

6.

The God Delusion

7.

Bernard Malamud

8.

Hermann Hesse

9.

The Scarlet Letter

10.

The Exorcist

11.

Martin Luther King Jr

12.

George Eliot

Bonus:

Living with the Gods: On Beliefs and Peoples is a book about religious belief through the millennia by Neil MacGregor, former director of both the National Gallery and the British Museum. Like MacGregor's earlier *A History of the World in 100 Objects*, the book accompanied a series on BBC Radio 4.

A7

QUIZ 8

ROUND 1

QUICKFIRE:
HOW FAR CAN YOU GO?

Books and transport . . .

1.

In a children's classic of 1908, which character is sentenced to 20 years in prison for motoring offences?

2.

Blanche DuBois is the main character of which 1947 play, set in New Orleans?

3.

Which vehicles are missing from the title of Marina Lewycka's bestselling debut novel of 2005: *A Short History of _____ in Ukrainian?*

Q8

4.

Who took ten years to sail back from the Trojan War to his home in Ithaca?

5.

Who wrote the 1957 novel *On the Road*, in which the main characters drive across America?

6.

In the 1970s Robert M. Pirsig scored a rare bestseller for a book about philosophy with *Zen and the Art of . . .* what?

7.

Which literary character consummates the relationship with her lover Léon in a carriage driving around Rouen?

8.

Which TV presenter's books include *Driven to Distraction*, *Round the Bend* and *Don't Stop Me Now*?

Q8

9.

The Alfred Hitchcock film *Strangers on a Train* is based on a novel by which writer, whose other books include *The Talented Mr Ripley*?

ANSWERS ON PP. 265–6

10.

The *Pequod* is the ship at the centre of which novel?

11.

Whose autobiographical account of the Arab revolt in the First World War, *Seven Pillars of Wisdom*, was first published in a general edition in 1935, soon after his death in a motorbike accident?

12.

Who caused controversy in 1973 with her feminist novel *Fear of Flying*?

Q8

247

ROUND 2

NAME THE AUTHOR

Can you guess the writer from these clues
(and, of course, the fewer you need the better)?

I.

A.   Three of his novels feature the university professor David Kepesh.

B.   His early fiction caused one rabbi to ask, 'What is being done to silence this man?'

C.   His second wife was the actress Claire Bloom.

D.   Nine of his novels feature the novelist Nathan Zuckerman.

Q8

ANSWERS ON PP. 267–9

2.

A.   He was the first cousin of the British prime minister Stanley Baldwin.

B.   He won the Nobel Prize in Literature in 1907 and, as of 2018, remains the youngest ever winner of the prize.

C.   One of his poems, containing the phrases 'those two impostors' and 'the unforgiving minute', is regularly voted the nation's favourite.

D.   The characters in one of his children's books include Shere Khan, Baloo and Mowgli.

3.

A.   She was born in Persia, now Iran.

B.   She grew up in Northern Rhodesia, now Zambia – where some of her books are set – before settling in Britain.

C.   She won the Nobel Prize in Literature in 2007 and, as of 2018, remains the oldest ever winner of the prize.

D.   Her books include *The Grass Is Singing*, the *Children of Violence* series and *The Golden Notebook*.

Q8

ANSWERS ON PP. 267–9

4.

A.   His first book – although not his biggest selling – was *A Syllabus of Plane Algebraic Geometry* (1860).

B.   One of his characters could believe 'six impossible things before breakfast'.

C.   The main character in his two most famous books was based on a real-life girl whose surname was Liddell.

D.   His own real name was Charles Dodgson.

5.

A.   A real song by U2 takes its lyrics from a fictional song in one of his books.

B.   When working in advertising, he wrote the line 'That'll do nicely' for American Express.

C.   Like the main character of his breakthrough novel, he was born in Bombay in 1947.

D.   For much of the 1990s he lived in hiding under the name Joseph Anton.

Q8

ANSWERS ON PP. 267–9

6.

A. In 2018 she joined the *Doctor Who* writing team, and co-wrote an episode in which the Doctor witnesses Rosa Parks not giving up her bus seat to a white man in 1950s Alabama.

B. She was the Children's Laureate between 2013 and 2015.

C. She chose the title of her breakthrough novel, she has said, on the grounds that it's 'one of those games that nobody ever plays after childhood, because nobody ever wins' . . .

D. . . . that title being *Noughts & Crosses*.

Q8

ANSWERS ON PP. 267–9

ROUND 3

EXTRACTS:
PLEASED TO MEET YOU (2)

*Can you identify the four celebrated fictional characters
making their first appearances here for one point – plus
the book and author in each case for another two?*

I.

*From a novel of 1887
(where the novel's name might be the hardest bit):*

This was a lofty chamber, lined and littered with count-less bottles. Broad, low tables were scattered about, which bristle with retorts, test-tubes, and little Bunsen lamps, with their blue flickering flames. There was only one stu-dent in the room, who was bending over a distant table absorbed in his work. At the sound of our steps he glanced round and sprang to his feet with a cry of pleasure. 'I've found it! I've found it,' he shouted to my companion, run-ning towards us with a test-tube in his hand. 'I have found

Q8

252

a re-agent which is precipitated by haemoglobin, and by nothing else.' Had he discovered a gold mine, greater delight could not have shone upon his features . . .

'How are you?' he said cordially, gripping my hand with a strength for which I should hardly have given him credit. 'You have been in Afghanistan, I perceive.'

'How on earth did you know that?' I asked in astonishment.

2.

From a novel of 1850:

I went in, and found there a stoutish, middle-aged person, in a brown surtout and black tights and shoes, with no more hair upon his head (which was a large one, and very shining) than there is upon an egg, and with a very extensive face, which he turned full upon me. His clothes were shabby, but he had an imposing shirt-collar on. He carried a jaunty sort of a stick, with a large pair of rusty tassels to it; and a quizzing-glass hung outside his coat, – for ornament, I afterwards found, as he very seldom looked through it, and couldn't see anything when he did.

'This,' said Mr Quinion, in allusion to myself, 'is he.'

'This,' said the stranger, with a certain condescending roll in his voice, and a certain indescribable air of doing something genteel, which impressed me very much, 'is Master Copperfield. I hope I see you well, sir?'

Q8

253

3.

From a novella of 1958 (although it may be more famous from the film version three years later):

I went out into the hall and leaned over the banister, just enough to see without being seen. She was still on the stairs, now she reached the landing, and the ragbag colours of her boy's hair, tawny streaks, strands of albino-blond and yellow, caught the hall light. It was a warm evening, nearly summer, and she wore a slim cool black dress, black sandals, a pearl choker. For all her chic thinness, she had an almost breakfast-cereal air of health, a soap and lemon cleanness, a rough pink darkening in the cheeks. Her mouth was large, her nose upturned. A pair of dark glasses blotted out her eyes. It was a face beyond childhood, yet this side of belonging to a woman. I thought her anywhere between sixteen and thirty; as it turned out, she was shy two months of her nineteenth birthday.

4.

From a novel of 1961, where this time the character you're after is just about to appear:

Along came Mary Macgregor, the last member of the set, whose fame rested on her being a silent lump, a nobody whom everybody could blame. With her was an outsider, Joyce Emily Hammond, the very rich girl, their delinquent, who had been recently sent to Blaine as a last hope, because no other school, no governess, could manage her. She still wore the green uniform of her old school. The

254

ANSWERS ON PP. 270–71

others wore deep violet. The most she had done, so far, was to throw paper pellets sometimes at the singing master. She insisted on the use of her two names, Joyce Emily. This Joyce Emily was trying very hard to get into the famous set, and thought the two names might establish her as a something, but there was no chance of it and she could not see why.

Joyce Emily said, 'There's a teacher coming out,' and nodded towards the gates . . .

Q 8

ANSWERS ON PP. 270–71

ROUND 4

ODD ONE OUT

In each of these batches of four,
what item doesn't belong and why?

I.

A. The author of the novels *Hotel World*, *The Accidental* and *How to Be Both*, all published between 2001 and 2014

B. The author of the poem 'Not Waving but Drowning'

C. The author of the 1948 novel *I Capture the Castle* – who later wrote a children's canine classic

D. The author of the 2018 novel *Never Greener* – who'd earlier co-written the TV comedy series *Gavin & Stacey*

Q8

ANSWERS ON PP. 272–4

2.

A. Ben Okri's novel about an African spirit child, which won the 1991 Booker Prize

B. The last novel in Pat Barker's *Regeneration* trilogy about the First World War, which won the 1995 Booker Prize

C. Monica Ali's first novel, which was shortlisted for the 2003 Man Booker Prize

D. Richard Flanagan's novel about the building of the Burma railway, which won the 2014 Man Booker Prize

(*Clue: the answer has nothing to do with the Booker Prize*)

Q8

ROUND 4

3.

A. The Yuletide story by Nikolai Gogol that shares its title with the words that follow ''Twas' in the first line of the most famous poem by Clement Clarke Moore

B. The collection in which the stories of Aladdin and Ali Baba first appeared

C. Frederick Forsyth's novel about an assassination attempt on General de Gaulle

D. The TV programme for young children on which a series of Ladybird books are based – among them *Igglepiggle's Lost Blanket* and *Sleep Tight, Upsy Daisy*

4.

A. The first novel by Caitlin Moran

B. The first novel by vlogger Zoe Sugg (aka Zoella), which was the fasting selling book of 2014

C. Anne Louvert, according to the title of a novel by Sebastian Faulks set in a French inn

D. The 2006 John Boyne novel set in a Nazi concentration camp

ANSWERS ON PP. 272–4

5.

A. The novels *White Fang* and *The Call of the Wild*

B. Martin Amis's fields in a novel of 1989

C. Peter Ackroyd's first novel, which shares its name with that of a historic event of 1666

D. In Shakespeare's *Romeo and Juliet*, Juliet's original suitor whom Romeo ends up killing beside her tomb

6.

A. The Henry James novel whose main character is Isabel Archer

B. The poet who said 'I awoke to find myself famous' after the success of his poem *Childe Harold's Pilgrimage*

C. Jim in the title of a 1900 novel by Joseph Conrad

D. Cedric Errol in an 1886 children's novel by Frances Hodgson Burnett, whose title lives on as a phrase meaning a spoiled or pampered child

Q 8

ANSWERS ON PP. 272–4

ROUND 5

QUICKFIRE:
THE SEA, THE SEA

Twelve aquatic questions (or aquatic answers) . . .

1.

In the poem by Edward Lear, which two creatures go to sea in 'a beautiful pea-green boat'?

2.

Which 1972 novel by Richard Adams features the characters Hazel, Bigwig and Fiver?

3.

What's the title of Jean Rhys's 1966 novel which tells the story of the first Mrs Rochester from *Jane Eyre*?

4.

In Greek mythology, what's the name of Jason's ship in his quest for the Golden Fleece?

5.

Which sea creature is missing from the title of this 2018 novel by Imogen Hermes Gowar: *The _____ and Mrs Hancock*?

6.

Captain Nemo is the captain of the submarine *Nautilus* in which 1870 novel by Jules Verne?

7.

His Joyful Water-Life and Death in the Country of the Two Rivers is the subtitle of a book by Henry Williamson – about an otter called what?

8.

John Grady Cole, Lacey Rawlins and Jimmy Blevins cross the Rio Grande into Mexico in whose 1992 novel *All the Pretty Horses*?

Q 8

ANSWERS ON PP. 275–6

9.

Who wrote the 1959 bestseller *Billy Liar*?

10.

Who wrote the 1863 bestseller *The Water Babies*?

11.

The poem that begins 'Razors pain you;/ Rivers are damp' and ends 'You might as well live' is by which writer – also known for her many one-liners delivered at a regular gathering of New York wits at the Algonquin Hotel in the 1920s?

12.

Which poem contains the lines: 'Water, water every where,/ Nor any drop to drink'?

ANSWERS
TO QUIZ 8

ROUND I

1.

Toad – in Kenneth Grahame's *The Wind in the Willows*

2.

A Streetcar Named Desire – by Tennessee Williams

3.

Tractors

4.

Odysseus – in Homer's *Odyssey*. (His Roman name of Ulysses is fine as an answer too.)

5.

Jack Kerouac

6.

Motorcycle Maintenance

7.

Emma/Madame Bovary – in Gustave Flaubert's *Madame Bovary*. (Within a year of the book coming out, in Hamburg you could hire cabs called Bovarys for sexual purposes.)

8.

Jeremy Clarkson

9.

Patricia Highsmith

10.

Moby-Dick – by Herman Melville

11.

T. E. Lawrence, aka Lawrence of Arabia. (The book had appeared in his lifetime but only in very limited or abridged editions.)

12.

Erica Jong. (I'll spare you the details of the controversy in a family quiz book.)

Bonus:

How Far Can You Go? is a 1976 novel by David Lodge about a group of Catholic students – and how they reacted to the changes in the Church during the 1960s.

ROUND 2

1.

Philip Roth, who fell foul of some of his fellow Jews for portraying his own background in a way that would provide 'fuel' for anti-Semites – not a criticism that the famously combative Roth took lying down. His marriage to Claire Bloom ended in some bitterness, with her later denouncing him in her 1996 autobiography *Leaving a Doll's House*. Coincidentally or not, Roth's first novel after her revelations was *I Married a Communist*, in which a mentally unstable actress publishes a book telling lies about her ex-husband.

2.

Rudyard Kipling, who was 41 when he also became the first ever British winner of the Nobel Prize in Literature. The poem is 'If–', with Triumph and Disaster as the two impostors that we should treat just the same – while also filling the unforgiving minute 'with sixty seconds' worth of distance run'. Shere Khan, Baloo and Mowgli appear in *The Jungle Book*.

A 8

3.

Doris Lessing, who was 11 days shy of her 88th birthday when she learned that she'd won the Nobel, after returning from shopping to find the press on her doorstep. When they told her the news, her response was: 'Oh Christ!'

4.

Lewis Carroll, author of the Alice books, whose day job was as a maths don at Christ Church, Oxford, where the dean was Henry Liddell, Alice's father. In 1864 Carroll gave Alice his own illustrated manuscript of *Alice's Adventures Under Ground* (the original title of *Alice's Adventures in Wonderland*) as a Christmas present. Needing money after the death of her husband, she sold it in 1928 to an American collector for £15,400 (around £1 million today).

5.

Salman Rushdie, whose breakthrough novel was *Midnight's Children*. (Unlike its main character Saleem Sinai, Rushdie was born only in the year of Indian independence, not at the precise moment of it.) The U2 song is 'The Ground Beneath Her Feet', with its lyrics taken from Rushdie's novel of the same name, about a global rock star. While working in advertising, he also – and perhaps more famously – came up with 'Naughty but nice' for cream cakes and 'Adorabubble', 'Delectabubble', 'Incredibubble' and so on for Aero chocolate. Joseph Anton was the name he took while in hiding following Ayatollah Khomeini's fatwa against him for writing *The Satanic Verses*.

6.

Malorie Blackman. *Noughts & Crosses*, which imagines a world in which the inequality between white and black people is reversed, was voted the nation's 61st favourite novel in the BBC's 'Big Read' project – meaning that it beat, among others, *A Tale of Two Cities*, *Lord of the Flies* and *Matilda*.

A 8

ROUND 3

1.

Sherlock Holmes in *A Study in Scarlet* by Arthur Conan Doyle. The narrator is, of course, Dr Watson, who has returned to London and is looking for somewhere to live. A friend of his happens to know that a man called Sherlock Holmes needs someone to share his lodgings and introduces them while Holmes is researching in a hospital laboratory. ('Holmes is a little too scientific for my tastes,' the friend warns Watson in advance. 'It approaches to cold-bloodedness.') For the record, Holmes makes his Afghanistan deduction from Watson's tan, haggardness, arm injury and military air.

2.

Mr Micawber in Charles Dickens's *David Copperfield*. Mr Quinion is David's employer at this point and is introducing him to the man who'll be his landlord – and one of literature's great scene-stealers.

3.

Holly Golightly in Truman Capote's *Breakfast at Tiffany's* – famously played in the film by Audrey Hepburn, where she had dark hair, although she did keep the black dress, pearl choker and sunglasses. Capote himself always wanted Marilyn Monroe to play Holly and apparently hated Hepburn in the part. The Holly of the film was also a sanitised version of the book's, with less drug-taking, lesbianism and quasi-prostitution.

4.

Miss Jean Brodie in *The Prime of Miss Jean Brodie* by Muriel Spark – 'Blaine' being the Marcia Blaine School for Girls, where Miss Brodie teaches and has her 'set' of favourite girls. In 1966 the book became a West End play starring Vanessa Redgrave, and three years later a film starring Maggie Smith, who won a Best Actress Oscar for her performance. Both the play and the film reduced the number of girls in the Brodie set from six to four, and poor Joyce Emily didn't make the cut in either.

A 8

ROUND 4

1.

The odd one out is **D** – because that's Ruth <u>Jones</u>. The others are all Smiths: A, Ali <u>Smith</u>; B, Stevie <u>Smith</u>; and C, Dodie <u>Smith</u>, whose later canine classic was *The Hundred and One Dalmatians*.

2.

The odd one out is **C** – Monica Ali's *Brick <u>Lane</u>*, as the others are all roads: Ben Okri's *The Famished <u>Road</u>*; Pat Barker's *The Ghost <u>Road</u>*; and Richard Flanagan's *The Narrow <u>Road</u> to the Deep North*.

3.

The odd one out is **C**, *The <u>Day</u> of the Jackal* – the others all having something of the night about them. Gogol's *The <u>Night</u> Before Christmas* – as it's called in the Penguin translation – is still a big favourite in Russia and Ukraine where it's traditionally read to children on Christmas Eve, even though it's quite a dark tale. Aladdin and Ali

A8

Baba first appeared in the *Arabian Nights* (aka *One Thousand and One Nights*), although these two stories weren't in the original anthology of Arabic folk tales that make up the work: they were either later discoveries or even inventions by Antoine Galland, the first Western translator of the *Arabian Nights*. And, on a less controversial note, the TV programme is *In the Night Garden . . .* – voiced (somewhat surprisingly) by Derek Jacobi – where Igglepiggle and Upsy Daisy are two of the characters.

4.

The odd one out is **D**, *The Boy in the Striped Pyjamas* – because the other three all contain 'girl' in the title: Moran's *How to Build a Girl*; *Girl Online* by Zoe Sugg; and Faulks's *The Girl at the Lion D'Or*.

5.

The odd one out is **D**: Count Paris, the others all being references to London. Jack London took part in the Klondike gold rush of the 1890s – an experience that left him with scurvy, malnutrition, several missing teeth, facial scarring, a dodgy abdomen and the raw material for the novels *White Fang* and *The Call of the Wild*. Martin Amis's *London Fields* was left off the shortlist for the 1989 Booker Prize – much to the regret of the chair of judges David Lodge – because the two female judges objected to its depiction of women. And Peter Ackroyd's debut novel was *The Great Fire of London*.

A 8

6.

The odd one out is **A**, as Isabel Archer's the heroine of James's *The Portrait of a Lady*. The other three, by contrast, are lords: the suddenly famous <u>Lord</u> Byron; Conrad's *<u>Lord</u> Jim*; and Little <u>Lord</u> Fauntleroy – which is what Cedric becomes when a lawyer visits his humble American home to reveal he's the rather improbable heir to a British earldom.

A8

ROUND 5

1.

The owl and the pussy-cat

2.

Watership Down – Hazel, Bigwig and Fiver being rabbits

3.

Wide Sargasso Sea

4.

The *Argo* (hence Jason and the Argonauts)

5.

Mermaid

A 8

6.

Twenty Thousand Leagues Under the Sea

7.

Tarka (*Tarka the Otter* is the book's main title)

8.

Cormac McCarthy

9.

Keith Waterhouse

10.

Charles Kingsley. (For the record, the full title is *The Water Babies: A Fairy Tale for a Land-Baby.*)

11.

Dorothy Parker

12.

The Rime of the Ancient Mariner – by Samuel Taylor Coleridge

A8

Bonus:

The Sea, the Sea is a novel by Iris Murdoch, which won the Booker Prize in 1978. She'd previously been shortlisted three times and would be shortlisted twice more.

QUIZ 9

ROUND I

All the answers here consist of two words that begin with consecutive letters of the alphabet . . .

I.

Whose bestselling works of history include *Stalingrad* and *Berlin: The Downfall 1945*?

2.

Whose first book was *The Voyage of the Beagle*?

3.

The musical that overtook *Cats* to become the longest running in West End history is based on (and named after) which novel of 1862?

Q9

279

4.

According to the title of a book by Christopher Isherwood, who changes trains?

5.

Who wrote the 1935 novel *Regency Buck*, the first of her many novels that were set in the Regency period?

6.

Whose travel books include *In Patagonia* and *The Songlines*?

7.

Which Australian-born writer's bestselling comic novels of the 1990s included *Foetal Attraction* and *Mad Cows*?

8.

Whose bestselling adventure stories of the post-war years included *The Lonely Skier* and *The Wreck of the Mary Deare*?

9.

Which pop singer's 2012 autobiography acknowledged that he'd had the same haircut for the past 45 years – which, he added, is what he had in common with the Queen?

Q9

10.

'Season of mists and mellow fruitfulness' is the first line of whose poem *To Autumn*?

11.

What's the only novel by Edith Wharton whose title is a man's name?

12.

Whose novels of the 2010s include *A Hologram for the King*, about the global financial crash, and *The Circle*, about the world's most powerful internet company?

ANSWERS ON PP. 297–8

ROUND 2

NAME THE AUTHOR

*Can you guess the writer from these clues
(and, of course, the fewer you need the better)?*

I.

A. His German wife was a distant cousin of the First World War flying ace known as the Red Baron.

B. His middle name was the same as H. G. Wells's first name.

C. In 1960 one of his novels sold around two million copies in the six weeks before Christmas – although by then he'd been dead for 30 years.

D. He was the son of a Nottinghamshire miner.

Q9

ANSWERS ON PP. 299–301

2.

A. She wrote the early 1970s sitcom *It's Awfully Bad for Your Eyes, Darling*, starring Joanna Lumley.

B. Her male characters include Basil Baddingham, Lysander Hawkley and Viking O'Neill.

C. Her first six novels have titles that are women's names.

D. Her later ones are set in the fictional county of Rutshire.

3.

A. Born in Calcutta, he was the frankly unsympathetic Irish expert for *Punch* magazine during the Great Famine.

B. One of his novels was filmed by Stanley Kubrick in 1975, resulting in what Martin Scorsese has called Kubrick's best movie.

C. Another of his novels – adapted several times for film and television – is named after a place in John Bunyan's *The Pilgrim's Progress*.

D. The main character of that novel is Becky Sharp.

Q9

4.

A. Her novels include *Delusions of Grandma* and *The Best Awful There Is*.

B. Her memoirs include *Wishful Drinking* and *Shockaholic*.

C. When she was two, her father left her mother for Elizabeth Taylor.

D. Her most famous film role was as Princess Leia.

5.

A. His first book was a biography of the Pre-Raphaelite painter and poet Dante Gabriel Rossetti.

B. In 1927 he married a woman with the same Christian name as his.

C. His oldest son became a journalist, a regular contributor to *Private Eye* and the editor of the *Literary Review*.

D. His novels include *Decline and Fall* and *Scoop*.

Q9

ANSWERS ON PP. 299–301

6.

A.   When he was 27, he married his 13-year-old cousin.

B.   His death, aged 40, in Baltimore remains a mystery to this day – and the theories as to its cause include murder, suicide, alcohol poisoning, carbon monoxide poisoning, syphilis, flu, rabies and a brain tumour.

C.   Arthur Conan Doyle credited him with creating the modern detective story.

D.   His works of fiction include 'The Murders in the Rue Morgue' and 'The Pit and the Pendulum'.

Q9

285

ANSWERS ON PP. 299–301

ROUND 3

EXTRACTS: PICTURE THIS

*The following passages are from novels that became films,
all of which were nominated for a Best Adapted Screenplay
Oscar. Can you name the novel in each case for one point,
and the author for another?*

I.

Oscar-nominated in 2008:

The crowds were bunching up again. In front of the canal bridge was a junction, and from the Dunkirk direction, on the road that ran along the canal, came a convoy of three-ton lorries which the military police were trying to direct into a field beyond where the horses were. But troops swarming across the road forced the convoy to a halt . . . There was a shout of, Take cover! And before anyone could even glance round, the mountain of uniforms was detonated. It began to snow tiny pieces of dark green serge.

Q9

2.

Oscar-nominated in 2016:

On the night before they were due to dock, she went to the dining room with Georgina, who told her that she looked wretched and that if she did not take care she would be stopped at Ellis Island and put in quarantine, or at least given a thorough medical examination. Back in the cabin, Eilis showed Georgina her passport and papers to prove to her that she would not have a problem entering the United States. She told her that she would be met by Father Flood.

3.

Oscar-nominated in 2007:

I put Sheba up for a week or so in my flat . . . How could I not? Sheba was so pitifully alone. It would have taken a very unfeeling individual to desert her. There is at least one more pre-trial hearing – possibly two – to be got through before the case goes before the Crown Court and, frankly, I don't think Sheba would make it on her own. Her barrister says that she could avoid going to the Crown Court altogether if she pleaded guilty to the charges. But Sheba won't hear of it. She regards a guilty plea . . . as unthinkable. 'There was no assault and I've done nothing indecent,' she likes to say.

Q9

ANSWERS ON PP. 302–3

4.

Oscar-nominated in 1997:

The sweat wis lashing oafay Sick Boy; he wis trembling. Ah wis jist sitting thair, focusing oan the telly, tryin no tae notice . . . He wis bringin me doon. Ah tried tae keep ma attention oan the Jean-Claude Van Damme video.

As happens in such movies, they started oaf wi an obligatory dramatic opening. Then the next phase ay the picture involved building up the tension through introducing the dastardly villain and sticking the weak plot thegither. Any minute now though, auld Jean-Claude's ready tae git doon tae some serious swedgin.

Q9

ANSWERS ON PP. 302–3

ROUND 4

ORDER, ORDER

*Can you place these three items in ascending,
correct or their usual order?*

1.

A. The Ken Kesey novel whose main character is
    Randle McMurphy, played by Jack Nicholson in
    the film version

B. The novel of 1844 whose main character is
    d'Artagnan

C. The novel that begins 'It was the best of times,
    it was the worst of times'

2.

A. John Webster's devil

B. Mark Haddon's house

C. Tom Sharpe's Porterhouse

Q9

ANSWERS ON PP. 304–6

ROUND 4

3.

A. Saul Bellow's adventurous Augie

B. H. E. Bates's first book about the Larkin family

C. Elizabeth von Arnim's novel *The Enchanted* _____.

4.

A. *The* _____ *Memoirs and Confessions of A Justified Sinner* (an 1824 novel by James Hogg that was voted number ten in a 2016 BBC Scotland poll to find the country's favourite Scottish novel)

B. The 1990s bestseller set on the island of Cephalonia

C. Barbara in the title of a play by George Bernard Shaw

5.

A. _____ *Song* (a 1932 novel by Lewis Grassic Gibbon that was voted number one in that same 2016 BBC Scotland poll

B. The author of *Dear Fatty* and *A Tiny Bit Marvellous*

C. Jonathan Pine in John le Carré's first post-Cold War novel

Q9

6.

A. The trees of Thika, according to the title of a 1959 memoir by Elspeth Huxley of her childhood in Kenya, which became a TV series in the 1980s

B. The Dan Brown novel that shares a title with Dante

C. The author of *A Far Cry from Kensington* and *The Ballad of Peckham Rye*

Q9

ANSWERS ON PP. 304–6

ROUND 5

QUICKFIRE:
THE SENSE OF AN ENDING

Each of the answers here begins and ends with the same letter.
(So if it was a football quiz, Celtic could be an answer,
because it begins and ends with 'c' – but Rangers couldn't.)

I.

What's the full name of the literary character who turns down a marriage proposal from Mr Elton – and later marries Mr Knightley?

2.

To Say Nothing of the Dog is the subtitle of which book by Jerome K. Jerome?

3.

Who wrote the *Alfie* series for children?

ANSWERS ON PP. 307–8

4.

Which Roald Dahl novel is narrated by the son of a poacher?

5.

The apocalyptic comic novel *Good Omens* was co-written by Terry Pratchett and which other author, whose books include *American Gods* and *The Ocean at the End of the Lane*?

6.

Who wrote the controversial 1955 novel that's narrated by Humbert Humbert?

7.

In the books by P. G. Wodehouse, what's the name of Bertie Wooster's least favourite aunt?

8.

What first name is shared by: the novelist and academic who edited the 1987 collection *The Penguin Book of Modern Short Stories*; the author of the 1947 novel *Under the Volcano*; and the author of the 2008 non-fiction book *Outliers*?

Q9

ANSWERS ON PP. 307–8

9.

The playwright Peter Shaffer – author of *Equus* and *Amadeus* – was born in which city?

10.

Isaac Bashevis Singer, the first Yiddish writer to win the Nobel Prize in Literature, grew up in which city?

11.

Which children's book series features the siblings John, Susan, Titty and Roger?

12.

In which Virginia Woolf novel does the main character change from a man to a woman and live for centuries?

Q9

ANSWERS ON PP. 307–8

ANSWERS
TO QUIZ 9

ROUND I

1.

Antony Beevor

2.

Charles Darwin

3.

Les Misérables – by Victor Hugo

4.

Mr Norris (that title being *Mr Norris Changes Trains*)

5.

Georgette Heyer

6.

Bruce Chatwin

A9

7.

Kathy Lette

8.

Hammond Innes

9.

Rod Stewart

10.

John Keats

11.

Ethan Frome

12.

Dave Eggers

Bonus:

The Couple Next Door is a 2016 bestseller by Shari Lapena – a classic psychological thriller of the kind where it turns out that we mightn't know the people close to us as well as we thought . . .

ROUND 2

1.

D. H. Lawrence – where the D and H stand for David Herbert (H. G. Wells was Herbert George). That huge pre-Christmas bestseller was *Lady Chatterley's Lover*, after Penguin won the celebrated court case to publish the full rude version in early November. Lawrence's wife Frieda's maiden name was the same as the Red Baron's – von Richthofen – and during the First World War this meant that the couple had to leave Cornwall, because their neighbours accused them of being German spies.

2.

Jilly Cooper, who began with the novels *Emily*, *Bella*, *Harriet*, *Octavia*, *Imogen* and *Prudence* – before publishing *Riders* in 1985, which introduced her long-running cad Rupert Campbell-Black, who Cooper has said was partly based on Andrew Parker-Bowles, the former husband of the Duchess of Cornwall.

A9

3.

William Makepeace Thackeray, whose *The Luck of Barry Lyndon* became Kubrick's *Barry Lyndon*. *Vanity Fair* is the one with the title taken from Bunyan and has been filmed at least seven times – including four silent movie versions – as well as being a regular choice of TV costume drama.

4.

Carrie Fisher, the daughter of the Hollywood actress Debbie Reynolds and Eddie Fisher, a 1950s teen idol whose relationship with Taylor caused a scandal that badly damaged his career. (Even so, he became Taylor's fourth husband – not bad going, given that she was 27 at the time.) Reynolds died the day after her daughter did, in 2016.

5.

Evelyn Waugh, whose first wife was Evelyn Gardner. (Friends referred to them as 'He-Evelyn' and 'She-Evelyn'.) That journalist son was Auberon, known as Bron, one of Waugh's seven children with his second wife Laura. 'The presence of my children affects me with deep weariness and depression,' Waugh characteristically recorded in his diary in December 1946. 'Bron is clumsy and dishevelled, sly, without intellectual, aesthetic or spiritual interest' – although on a happier note, his old dad later noted that Bron 'is lazy but not very stupid'.

6.

Edgar Allan Poe, who was found wandering deliriously around Baltimore wearing somebody else's clothes in October 1849. He was then taken to hospital, where he died four days later – without ever recovering enough to explain what had happened. Speaking at a dinner to mark the centenary of Poe's birth in 1909, Doyle said, 'Where was the detective story until Poe breathed life into it?' – as well he might, given that Poe created such staples of the genre as the sealed-room mystery, the stories being told by the detective's admiring friend and the detective's all-round deductive brilliance. When Dr Watson first gets to know Holmes, he says to him, 'You remind me of Edgar Allan Poe's Dupin' – although Holmes naturally replies, 'In my opinion, Dupin was a very inferior fellow.'

A9

ROUND 3

I.

Atonement by Ian McEwan, a novel that ranges from a country house in the 1930s through the Second World War to London in 1999. It also plays games with its readers that I'd better not reveal in case you haven't been one of them yet. The film featured a famous five-minute tracking shot of the beach at Dunkirk – which was filmed in Redcar. The screenwriter was Christopher Hampton, who'd previously won a Best Adapted Screenplay Oscar for *Les Liaisons Dangereuses*.

2.

Brooklyn by Colm Tóibín, where the screenplay was written by Nick Hornby. The main character is Eilis Lacey – played in the film by Saoirse Ronan – who in the 1950s moves to Brooklyn from Enniscorthy, County Wexford, where Tóibín was born and where many of his novels are set.

3.

Notes on a Scandal by Zoë Heller, about a secondary-school teacher who has an affair with a fifteen-year-old student – and is forced to rely on her friend Barbara, who narrates the novel and is not as friendly as she seems. Sheba and Barbara were played in the film by Cate Blanchett and Judi Dench, both of whom were Oscar-nominated too, along with the film's writer Patrick Marber, actor, director, playwright and co-creator of Alan Partridge.

4.

The opening of *Trainspotting* by Irvine Welsh, featuring a group of Edinburgh heroin addicts, before the book hits its full sweary stride – although that ellipsis was definitely required in a family quiz book. The characters in the book, Welsh's first, reappear in his later novels *Skagboys*, *Porno*, *The Blade Artists* and *Dead Man's Trousers*.

A9

ROUND 4

1.

A, C, B – one, two, three:

<u>One</u> Flew Over the Cuckoo's Nest

A Tale of <u>Two</u> Cities (Charles Dickens)

The <u>Three</u> Musketeers (Alexandre Dumas)

2.

B, A, C – red, white and blue:

The <u>Red</u> House (a 2012 novel)

The <u>White</u> Devil
(Webster's famously grisly Jacobean revenge tragedy)

Porterhouse <u>Blue</u>
(Tom Sharpe's comic novel of 1974, which became a
successful TV series in the 1980s)

3.

A, C, B – consecutive months:

The Adventures of Augie <u>March</u>
(Bellow's breakthrough novel of 1953)

The Enchanted <u>April</u>
(the award-winning 1991 film version dropped the 'The')

The Darling Buds of <u>May</u>
(which began a series of novels that were turned into the
hugely popular TV drama of the 1990s)

4.

A, B, C – ascending order of army ranks:

The <u>Private</u> Memoirs and Confessions of a Justified Sinner

<u>Captain</u> Corelli's Mandolin (Louis de Bernières)

<u>Major</u> Barbara (about an idealistic Salvation Army officer)

5.

B, A, C – times of the day:

<u>Dawn</u> French

<u>Sunset</u> Song (set in the tough rural Scotland of
the early twentieth century)

The <u>Night</u> Manager

A9

6.

C, A, B – the progress of a fire:

Muriel <u>Spark</u>

The <u>Flame</u> Trees of Thika

<u>Inferno</u>

ROUND 5

1.

Emma Woodhouse (in Jane Austen's *Emma*)

2.

Three Men in a Boat

3.

Shirley Hughes

4.

Danny, the Champion of the World

5.

Neil Gaiman

6.

Vladimir Nabokov (the controversial novel being *Lolita*)

A9

7.

Agatha

8.

Malcolm (Bradbury, Lowry and Gladwell respectively)

9.

Liverpool

10.

Warsaw

11.

Swallows and Amazons – by Arthur Ransome

12.

Orlando

Bonus:

The Sense of an Ending is a novel by Julian Barnes about – among other things – the unreliability of memory. The book won the 2011 Man Booker Prize, a prize Barnes had previously described as 'posh bingo'. After his victory, Barnes maintained that it's still best to treat the Booker as 'a lottery' – until you win, 'when you realise that the judges are the wisest heads in literary Christendom'.

QUIZ 10

ROUND I

RODERICK RANDOM

Twelve questions with no theme at all . . .

I.

The first edition of which book, first published in 1852 and never out of print since, began: 'Ens, entity, being, existence'?

2.

A Woman's Work is the 2017 autobiography of which politician, who twice served as the acting leader of the Labour Party?

3.

A Brief History of Humankind is the subtitle of which 2014 book by Yuval Noah Harari?

QIO

ANSWERS ON PP. 329–30

4.

Which prophecy became a bestseller in the 1990s for James Redfield?

5.

Which art became a bestseller in the 1980s for Donald Trump?

6.

Whose first two novels, in the 1990s, were *Watermelon* and *Lucy Sullivan Is Getting Married*?

7.

Dude, Where's My Country? and *Stupid White Men* are among the books by which alliteratively named film-maker and author?

8.

Which alliteratively named novelist was born in South Shields in 1906, left school at fourteen to go into domestic service and ended up as one of Britain's bestselling writers of the twentieth century?

9.

Who wrote *The Power*, which won the 2017 Bailey's Women's Prize for Fiction – and was named by Barack Obama as one of his favourite books of that same year?

Q10

10.

In 2002, which author – whose middle name is Kroeber – became the first woman to receive the annual Grand Master Award from the Science Fiction and Fantasy Writers of America?

11.

Which Indian writer, whose surname is a male first name, won the 1997 Booker Prize with her first novel *The God of Small Things*?

12.

Which Irish writer, whose surname is a male first name, was shortlisted for the Booker Prize four times between 1970 and 2002 – when his shortlisted book was *The Story of Lucy Gault*?

ROUND 2

NAME THE AUTHOR
CELEBRITY SPECIAL

These six writers are better known for other things – but, as ever, the fewer clues you need, the better . . .

I.

A. For a time, she was known to certain people as 'Renaissance'.

B. Her maiden name was Robinson.

C. In 1991, while a practising lawyer, she became an assistant to the Mayor of Chicago.

D. Between January 2009 and January 2017 her address was 1600 Pennsylvania Avenue, Washington, DC.

2.

A. His first TV role, in 1964, was as Bert Bradshaw in the soap opera *Crossroads*.

B. In 2006 he was voted British TV's 'greatest ever star' in a public poll that was part of ITV's 50th anniversary celebrations.

C. One of his long-running TV acting roles was as a detective created by R. D. Wingfield.

D. Another was as a character who in 2018 was voted by the British public as having provided TV's greatest ever moment – when he fell through a bar . . .

3.

A. Her father was the Mayor of Bath.

B. In 2017 she was voted Best TV Judge at the National Television Awards.

C. Her 2013 autobiography was called *Recipe for Life*.

D. In 2012 she received a CBE for 'services to the culinary arts'.

Q10

ANSWERS ON PP. 331–3

4.

A. He has the same initials as the author of the 1776 pamphlet *Common Sense*, which advocated American independence from Britain.

B. After leaving Sandhurst in 1992, he joined the Army Air Corps.

C. Between December 2015 and June 2016 he grew 2 to 3cm, even though he was in his forties at the time.

D. In 2016 he became the first Briton to be honoured by the Queen while not on Earth.

5.

A. In the sitcom *Friends*, she played the stepmother of Emily, the English woman Ross briefly married.

B. She wrote the Spice Girls jukebox musical *Viva Forever!*

C. She was the voice of the evil fairy godmother in *Shrek 2*.

D. Her TV roles include Caroline Martin and Edina Monsoon.

NAME THE AUTHOR CELEBRITY SPECIAL

6.

A. He co-founded Eton's first mountaineering society.

B. He was christened Edward, but the nickname that his older sister gave him when he was a week old has stuck ever since.

C. In 2009, aged 35, he became the UK's youngest ever Chief Scout.

D. His TV shows include *The Island* and *Born Survivor*.

ANSWERS ON PP. 331–3

ROUND 3

EXTRACTS:
HOW SOME OF IT ENDED

Can you name the books that end like this – and who wrote them? As you can probably guess by now, it's one point for the book and one for the author.

I.

From a novel of 1890:

When they entered, they found hanging upon the wall a splendid portrait of their master as they had last seen him, in all the wonder of his exquisite youth and beauty. Lying on the floor was a dead man, in evening dress, with a knife in his heart. He was withered, wrinkled, and loathsome of visage. It was not till they had examined the rings that they recognized who it was.

2.

From a novel of 1906:

Now the house door opens. Bobbie's voice calls:–

'Come in, Daddy; come in!'

He goes in and the door is shut. I think we will not open the door or follow him. I think that just now we are not wanted there. I think it will be best for us to go quickly and quietly away. At the end of the field, among the thin gold spikes of grass and the harebells and Gipsy roses and St John's Wort, we may just take one last look, over our shoulders, at the white house where neither we nor anyone else is wanted now.

3.

From a novel of 1897:

I took the papers from the safe where they had been ever since our return so long ago. We were struck with the fact, that in all the mass of material of which the record is composed, there is hardly one authentic document; nothing but a mass of typewriting, except the later note-books of Mina and Seward and myself, and Van Helsing's memorandum. We could hardly ask any one, even did we wish to, to accept these as proofs of so wild a story. Van Helsing summed it all up as he said, with our boy on his knee:–

'We want no proofs; we ask none to believe us! This boy will some day know what a brave and gallant woman his mother is. Already he knows her sweetness and loving care; later on he will understand how some men so loved her, that they did dare much for her sake.'

JONATHAN HARKER

319

ANSWERS ON PP. 334–5

4.

From a children's book of 1928:

So they went off together. But wherever they go, and whatever happens to them on the way, in that enchanted place on the top of the Forest a little boy and his Bear will always be playing.

ANSWERS ON PP. 334–5

ROUND 4

CONNECT THREE

Can you link the three items in each case?

1.

The author of the 2019 picture book *Swarm of Bees* – who'd earlier made his name with the thirteen-book sequence *A Series of Unfortunate Events*

Anthony Burgess's 1962 novel that became a controversial film

The black marketeer in Graham Greene's *The Third Man*, played in the film version by Orson Welles

321

ANSWERS ON PP. 336–7

2.

The 1970s bestseller by Jack Higgins, about a German wartime plot to kidnap Churchill, that takes its title from a phrase broadcast from the first Moon landing

The Barry Hines novel of 1968 that became a Ken Loach film

Helen Macdonald's 2014 Costa Book of the Year and winner of the Samuel Johnson Prize (now the Baillie Gifford Prize for Non-Fiction)

3.

The Playboy of the Western World

Disgrace

Peter Pan

4.

The author of the 2018 novel *Mad Blood Stirring* – who's also been a presenter of *The Radio 1 Breakfast Show* and the Radio 2 *Drivetime* show

The author of *In the Psychiatrist's Chair*, based on the interviews he conducted in his Radio 4 series of the same name

Ruth Rendell's detective, who first appeared in her first novel *From Doon with Death*

ANSWERS ON PP. 336–7

5.

Pal ____, a novel by John O'Hara that became a Rodgers and Hart musical

The Life and Times of the Thunderbolt ____, a memoir of childhood by Bill Bryson

Brother and sister Charles and Mary, whose *Tales from Shakespeare* has never been out of print since it was published in 1807

6.

Call the Midwife

The Story of Tracy Beaker

The Day of the Triffids

ANSWERS ON PP. 336–7

ROUND 5

JOURNEY'S END

And so, farewell . . .

I.

Goodbye Soldier was the sixth volume of whose war memoirs that began with *Adolf Hitler: My Part in His Downfall*?

2.

Whose last novel was *Finnegans Wake*?

3.

Titus Alone is the last book in which trilogy by Mervyn Peake?

4.

What's the first line of the poem whose last line is 'And dances with the daffodils'?

ANSWERS ON PP. 338–40

5.

Holiday at the Dew Drop Inn was the third and final novel in the trilogy that began with *The Family from One End Street* – by which British children's writer?

6.

Moominvalley in November was the last novel in the Moomins series – by which Finnish children's writer?

7.

Dylan Thomas and Albert Einstein both died in which decade?

8.

Which First World War poet, whose poems include 'Anthem for Doomed Youth' and 'Dulce et Decorum Est', was killed in action a week before the Armistice was signed?

9.

Who wrote the 1995 novel *The Moor's Last Sigh*?

ANSWERS ON PP. 338–40

10.

Doug Naylor's novel *Last Human*, whose main character is Dave Lister, is based on which TV series that Naylor also wrote – initially with Rob Grant?

11.

What was the title of Robert Graves's 1929 autobiography – a title that, he rather ruefully acknowledged, was 'my sole contribution to Bartlett's *Dictionary of Familiar Quotations*'?

12.

Atlas Shrugged was which writer's last novel?

ANSWERS
TO QUIZ 10

ROUND I

I.

Roget's Thesaurus

2.

Harriet Harman

3.

Sapiens

4.

The Celestine Prophecy

5.

The Art of the Deal

6.

Marian Keyes

7.

Michael Moore

8.

Catherine Cookson

9.

Naomi Alderman

10.

Ursula K. Le Guin

11.

Arundhati Roy

12.

William Trevor

Bonus:

Roderick Random is a 1748 novel by the Scottish author Tobias Smollett – written, as he explained in the Preface, to arouse 'generous indignation . . . against the vicious disposition of the world'. The book was partly based on Smollett's own experiences as a naval surgeon.

A 10

ROUND 2

1.

Michelle Obama, author of the bestselling 2018 autobiography *Becoming*, which chronicles her life before, during and after her time as First Lady, including her many projects to help young women. (She also spills a few beans on her marriage to Barack.) Renaissance was her Secret Service code name – alongside Barack's Renegade – and 1600 Pennsylvania Avenue is the address of the White House.

2.

David Jason, whose bestselling 2013 autobiography *My Life* was followed four years later by *Only Fools and Stories*. R. D. Wingfield created Inspector 'Jack' Frost, played by Jason between 1992 and 2010 in *A Touch of Frost*. And, as you possibly know, it was Del Boy Trotter from *Only Fools and Horses* who fell through that bar.

A I O

3.

Mary Berry, author of more than 75 cookery books as well as that autobiography. Her Best Judge Award for *The Great British Bake Off*, voted for by the public, came after she'd announced her departure from the show following its move from the BBC to Channel 4 – and Simon Cowell and David Walliams were among her defeated rivals. (No wonder she punched the air in a way well worth looking up on the internet.)

4.

Tim Peake, whose book *Ask an Astronaut* details his time on the International Space Station, where in the absence of gravity he grew those extra centimetres – and where he became a Companion of the Order of St Michael and St George shortly before his return to Earth. The hugely influential *Common Sense* is by Thomas Paine, who later moved to Paris and became part of the French Revolution too.

5.

Jennifer Saunders, whose autobiography *Bonkers: My Life in Laughs* includes the story of her wedding to Ade Edmondson, when the vicar was so star-struck by the celebs in the congregation that he kept getting distracted from the business in hand. She played Caroline Martin in her Women's Institute sitcom *Jam & Jerusalem* and Edina 'Eddie' Monsoon in *Absolutely Fabulous*. (*Viva Forever!*, I think it's fair to say, was not her greatest success.)

A 10

6.

Bear Grylls. His first book *Facing Up* was about his ascent of Everest aged 23 – and 18 months after breaking three vertebrae in a parachuting accident. It's since been followed by around 20 more, including the autobiography *Mud, Sweat and Tears*, several survival guides, accounts of his various adventures and *Mission Survival*, a fiction series for younger readers.

ROUND 3

1.

From *The Picture of Dorian Gray* by Oscar Wilde, his only novel, in which Dorian famously stays young and beautiful despite his dissolute ways – with the portrait in his attic doing the ageing for him. Until the final scene, that is, when Dorian's attempt to destroy the painting with a knife mysteriously leaves him fatally stabbed instead.

2.

From *The Railway Children* by E. Nesbit: the coda that follows the celebrated scene of Bobbie crying 'Oh! My Daddy, my Daddy' when she sees her father coming up the platform after his release from his false imprisonment on spying charges. (The 1970 film version omitted the 'Oh' and the first 'my'.)

A IO

3.

From *Dracula* by Bram Stoker: the note added seven years after the main events of the novel by Jonathan Harker, whose visit to Transylvania set the action of the novel in motion. Mina is his wife and Seward was a student of the book's main vampire expert Dr Van Helsing. *Dracula* was not a huge seller in Stoker's lifetime, but luckily it had a longer afterlife than his first book, *The Duties of Clerks of Petty Sessions in Ireland* (1879), a legal administration handbook.

4.

From *The House at Pooh Corner* (note: not *Winnie-the-Pooh*) by A. A. Milne, the second of the two Pooh books and the one that introduced Tigger.

ROUND 4

1.

Citrus fruits: <u>Lemon</u>y Snicket wrote *Swarm of Bees*; Burgess's *A Clockwork <u>Orange</u>* was turned into the Stanley Kubrick film that Kubrick himself withdrew from release in Britain after it had apparently inspired copycat violence; Orson Welles played Greene's Harry <u>Lime</u>.

2.

Birds of Prey: Higgins's *The <u>Eagle</u> Has Landed*; Hines's *A <u>Kestrel</u> for a Knave* (which became the film *Kes*); Macdonald's *H Is for <u>Hawk</u>*.

3.

Authors who used the initials J. M.: <u>J. M.</u> Synge's play *The Playboy of the Western World* caused riots when it was first performed in Dublin in 1907 – by republicans and nationalists who considered Synge's realism an insult to Ireland (and especially to Irish womanhood); <u>J. M.</u> Coetzee's *Disgrace* won the 1999 Booker Prize; <u>J. M.</u> Barrie created Peter Pan.

4.

Irish counties: Simon <u>Mayo</u>; Anthony <u>Clare</u>; Inspector <u>Wexford</u>.

5.

Young animals: *Pal Joey*; *The Life and Times of the Thunderbolt Kid*; Charles and Mary <u>Lamb</u> (also known for the less happy fact that in 1796 Mary stabbed their mother to death during a mental breakdown – Charles then looked after her until his death in 1834).

6.

Books by authors with the initials J. W.: *Call the Midwife* is by Jennifer Worth – and, as you may know, inspired a TV drama series; Jacqueline Wilson wrote *The Story of Tracy Beaker*; and *The Day of the Triffids* is a 1951 novel by John Wyndham about killer plants.

AIO

ROUND 5

1.

Spike Milligan

2.

James Joyce

3.

Gormenghast

4.

'I wandered lonely as a cloud' – by William Wordsworth

5.

Eve Garnett

6.

Tove Jansson

7.

The 1950s: Thomas in 1953, Einstein in 1955. (Einstein qualifies for this book because of Penguin's *The Essential Einstein*, edited by Stephen Hawking.)

8.

Wilfred Owen

9.

Salman Rushdie

10.

Red Dwarf

11.

Goodbye to All That

12.

Ayn Rand. (*Atlas Shrugged* was the novel that followed *The Fountainhead*, as featured in the title of the very first round in this book – and so, rather neatly I like to think, brings us back to where we started all those quizzes ago.)

A10

Bonus:

Journey's End is a play of 1928 by R. C. Sherriff set in the British trenches during the First World War. Although it's been regularly revived ever since, Sherriff initially had trouble finding a theatre that would stage it. Several producers told him that people didn't want dramas about the war, and one said, 'How can I put on a play with no leading lady?' – which is why *No Leading Lady* was the title of Sherriff's 1967 autobiography.

Acknowledgements

As you can probably tell, the introduction is indebted to two books in particular: Alan Connor's *The Joy of Quiz* (Particular Books, 2016) and Jeremy Lewis's *Penguin Special: The Life and Times of Allen Lane* (Viking, 2005) – both now available as Penguin paperbacks. Also useful were *The Penguin Classics Book* by Henry Eliot (Penguin, 2018), *The Book of Penguin* (2009) and *Question Time: A Journey Round Britain's Quizzes* (Weidenfeld and Nicolson, 2017) by Mark Mason.

The round Name the Author is based on a monthly quiz I write for *Reader's Digest* – many thanks to the editors Anna Walker and Eva Mackevic for allowing me to expand it here.

Thanks are also due to John Walsh and Sebastian Faulks – the team captains on the BBC Radio 4 books quiz *The Write Stuff* which I wrote and hosted for seventeen thoroughly enjoyable series. Together with the show's guests and producers, they greatly helped to sharpen my quizzing wits and keep me on my quizzing toes. Coming from an extended family of quiz-lovers also helped, mind you – so huge thanks as well to all the Waltons, Doyles, Walton-Doyles and McNamaras for lots of quizzing fun over the years.

Finally, thanks to Connor Brown and Emma Horton for their thoughtful and skilful editing.

Text permissions

He just wanted a decent book to read ...

Not too much to ask, is it? It was in 1935 when Allen Lane, Managing Director of Bodley Head Publishers, stood on a platform at Exeter railway station looking for something good to read on his journey back to London. His choice was limited to popular magazines and poor-quality paperbacks – the same choice faced every day by the vast majority of readers, few of whom could afford hardbacks. Lane's disappointment and subsequent anger at the range of books generally available led him to found a company – and change the world.

'We believed in the existence in this country of a vast reading public for intelligent books at a low price, and staked everything on it'
Sir Allen Lane, 1902–1970, founder of Penguin Books

The quality paperback had arrived – and not just in bookshops. Lane was adamant that his Penguins should appear in chain stores and tobacconists, and should cost no more than a packet of cigarettes.

Reading habits (and cigarette prices) have changed since 1935, but Penguin still believes in publishing the best books for everybody to enjoy. We still believe that good design costs no more than bad design, and we still believe that quality books published passionately and responsibly make the world a better place.

So wherever you see the little bird – whether it's on a piece of prize-winning literary fiction or a celebrity autobiography, political tour de force or historical masterpiece, a serial-killer thriller, reference book, world classic or a piece of pure escapism – you can bet that it represents the very best that the genre has to offer.

Whatever you like to read – trust Penguin.